ENERGY SCIENCE, ENGINEERING A

ENERGY SAVING AND STORAGE IN RESIDENTIAL BUILDINGS

ENERGY SCIENCE, ENGINEERING AND TECHNOLOGY

Additional books in this series can be found on Nova's website under the Series tab.

Additional E-books in this series can be found on Nova's website under the E-books tab.

ENERGY POLICIES, POLITICS AND PRICES

Additional books in this series can be found on Nova's website under the Series tab.

Additional E-books in this series can be found on Nova's website under the E-books tab.

ENERGY SCIENCE, ENGINEERING AND TECHNOLOGY

ENERGY SAVING AND STORAGE IN RESIDENTIAL BUILDINGS

Alicja Siuta-Olcha
and
Tomasz Cholewa

Nova Science Publishers, Inc.
New York

For permission to use material from this book please contact us:
Telephone 631-231-7269; Fax 631-231-8175
Web Site: http://www.novapublishers.com

Additional color graphics may be available in the e-book version of this book.

Library of Congress Cataloging-in-Publication Data

Energy saving and storage in residential buildings / editors, Alicja Siuta-Olcha and Tomasz Cholewa.
p. cm.
Includes bibliographical references and index.
ISBN 978-1-62100-167-6 (softcover)
1. Dwellings--Energy conservation. 2. Dwellings--Energy conservation--Europe. 3. Solar energy--Passive systems. 4. Solar energy--Passive systems--Europe. I. Siuta-Olcha, Alicja. II. Cholewa, Tomasz.
TJ163.5.D86E5285 2011696--dc23

2011035432

Published by Nova Science Publishers, Inc. † New York

CONTENTS

ABSTRACT

The residential sector consumes a vast amount of energy. This situation has resulted in an increased interest by the scientific community in the subject of energy saving. In the present work, we examine the characteristics of some newer systems and options for energy savings in residential buildings. Sources for optimally heating buildings, energy systems distribution, thermal energy storage systems, control systems among other improvements and operations, which may help to reduce energy consumption in the residential sector, are discussed The possibilities of use of heat pumps in technical systems, passive solar heating and water heating in domestic hot-water systems, are outlined. Special attention is paid to the problem of energy storage in energy systems. The varying behaviour of the thermal conditions in a solar water storage tank is analysed. Besides the technical possibilities resulting in the diminution of energy consumption in the residential sector, emphasis is also placed on the education of the final energy consumers. The modernization of existing buildings and their energy distribution systems, the goal of which is the adaptation of highly energy-consuming buildings to energy-saving standards, is also examined.

Chapter 1

INTRODUCTION

The major part of the world's energy-needs is still covered by fossil fuels, such as oil, coal and gas. Unfortunately, known reserves of these fuels, though increased, for example, in the case of gas thanks to discoveries of new deposits (Figure 1), are limited and will, in the long run, become exhausted. It is estimated [1] that oil will be sufficient for 35, coal for 107 and gas for 37 years.

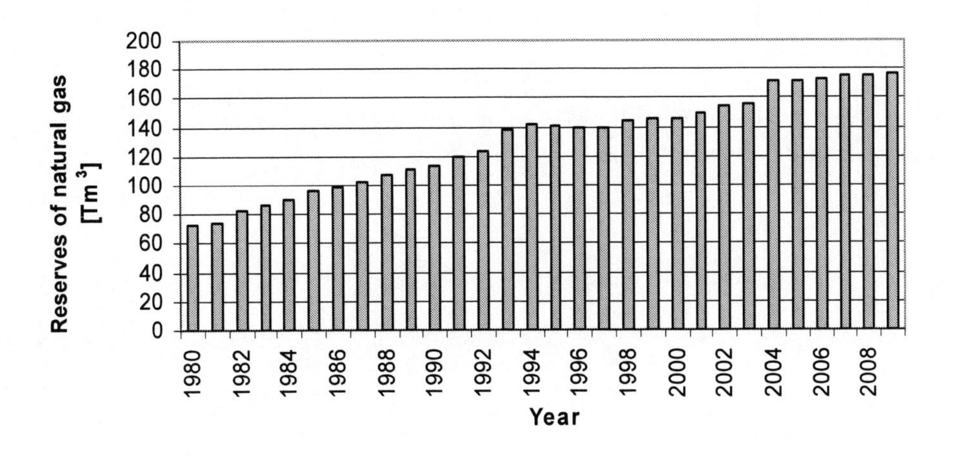

Figure 1. Reserves of natural gas in the world [2].

According to other sources, conventional oil reserve estimates are (in billion barrels): 1210 [3-5] and 1120 [5].

Another problem, connected with energy demand, is global warming, which is caused by greenhouse-gases. Among such gases are water vapour, carbon dioxide, ozone, methane and nitrogen oxides. It is known that the natural proportions of greenhouse-gases are augmented by vast amounts of pollution released to the atmosphere from global economic development and its associated fossil fuel combustion.

Bearing this in mind, the limitation of energy consumption and the growth of energy use from renewable energy sources, has become a major incentive. To advance this goal the European Union introduced an energy-climatic package, defining the following so-called „3x20" targets:

- an increase in energy-efficiency of about 20% by the year 2020,
- increased participation of renewable energy sources to 20% of the entire final energy consumption in the EU by the year 2020 and an increase to 10% of biofuels in transportive fuels,
- diminution of greenhouse-gases emission, of at least 20% in comparison to 1990, and even to 30%, on the condition that other developing countries will commit themselves to a comparable reduction of these emissions. Selected developing countries will also make appropriate contributions commensurate with their possibilities.

Special attention is paid to increasing the energy-efficiency in the residential sector, as it uses, on average ca. 30% of energy worldwide [6]. In the EU, buildings consume about 40% of the final energy [7,8]. This energy is used particularly for heating and hot water preparation and it is responsible for 30% of carbon dioxide emissions [9].

Chapter 2

THE RESIDENTIAL SECTOR IN EUROPE

Regarding the residential sector, which uses about 40% of the total energy in Poland, it is relevant to consider the following major usage sectors:

- Space heating (SH) and space cooling (SC) - energy required to support thermal losses incurred across the building envelope due to conduction and radiation, as well as air infiltration/ventilation in an effort to maintain a comfortable temperature and air quality within the living space;
- Domestic hot water (DHW) - energy required to heat water to an appropriate temperature for occupant and appliances (washing machine etc.);
- Appliances and lighting (AL) - energy consumed to operate common appliances (e.g. refrigerator and coffee maker) and for the provision of adequate lighting.

Among these, most of the energy is consumed by heating systems not only in Poland but also in other EU countries (Figure 2). For this reason, activities focusing on the effective use of energy should concentrate especially on this aspect.

Besides the diminution of the role of heating in the total energy balance of a single flat, comparing years 1993 and 2002 (from 73.1% to 71.2%) in Poland and from 77% (in 1997) to 75% (in 2007) in EU27 (the twenty seven members of the current European Union), may be noticed. This was caused by increased thermal isolation of new buildings, thermal renovation activities as well as an enlargement of electric energy consumption in flats.

The modern trend to acquire "labour-saving" and other devices using electricity has caused an increase in the proportion of electrical energy in the energy balance of residences.

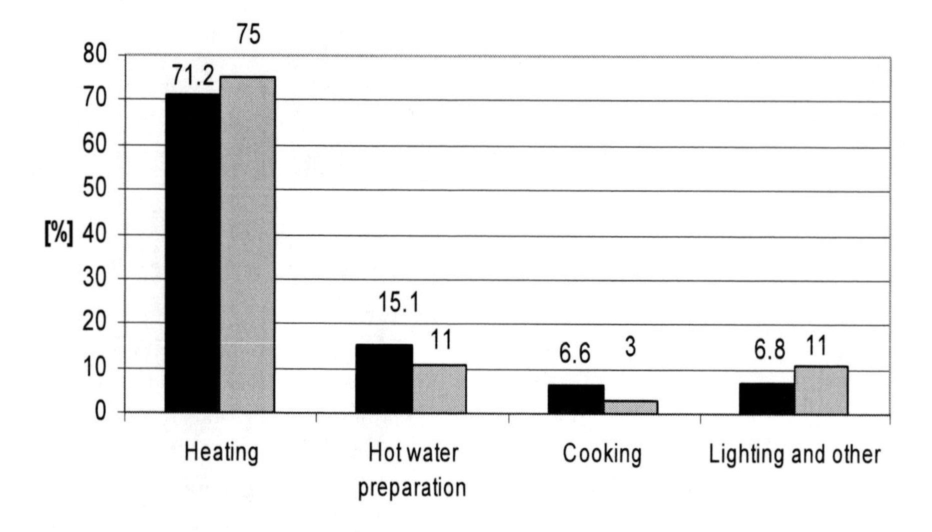

Figure 2. Energy balance in Polish and EU27 households [10,11].

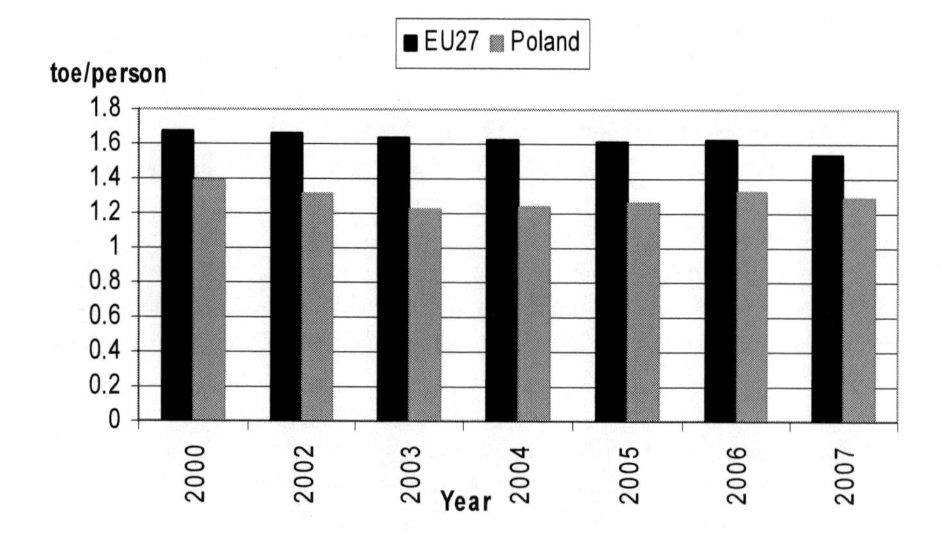

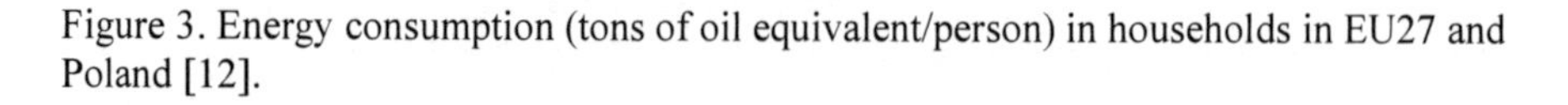

Figure 3. Energy consumption (tons of oil equivalent/person) in households in EU27 and Poland [12].

Among other causes, the growth of the electrical energy consumption can be mainly explained by:

- the high electricity consumption involved in the use of air-conditioning and mechanical ventilation devices in apartments;
- the use of low efficiency halogen lighting, as well as the enlargement of the lighting level of flats and houses;
- illumination of external spaces for decorative/ aesthetic reasons and safety;
- the large increase in energy consumption by devices such as large-screen television sets, stereo equipment, as well as their increased presence in the home;
- the increased energy consumption by devices left in stand-by status (building protection, fire-fighting systems, television, DVD players).

Research [13,14] has shown that the energy consumption by devices in the home in stand-by mode could be as high as 8÷12%.

Over recent years, in the EU (EU27) as well as in Poland (Figure 3), a diminution in the energy consumption of apartments/houses has been noted. This is partly ascribable to the intensified efforts at energy savings in these countries, but also to the ever stricter requirements on the thermal protection of buildings.

Among others, the following measures played major roles in the improvement of energy-efficiency in the Polish residential sector:

- introduction of an energy labeling system for buildings,
- the implementation of a thermal renovation fund,
- promotion of rational energy use in households.

The introduction of the energy labeling system for buildings, (Directive 2002/91/EC) [15] has led to improvements in the energy-efficiency of new and existing buildings, promotion of the use of renewable energy sources, increased safety of energy supply [16] and a reduction of greenhouse gas emissions [17].

During the preparation of the energy certificate for a building the building's owner is informed about activities which should be undertaken to achieve rational energy use. However, the European Standard *EN 15217* [18] has now been developed, and it describes methods for expressing energy efficiency and building certification. This directive has been more successful than the earlier European Council Directive 93/76/CEE [19], which presented energy certification as one of the cornerstones for achieving energy efficiency in buildings. However, being non-mandatory and furthermore riddled with ambiguities, it was not particularly successful throughout member states [20].

Another measure, which contributed to a considerable degree to the improvement of the energy-efficiency in Poland, was the thermal renovation fund. This fund has supported thermal renovation activities since as early as 1998. Help from the Fund embraces: the improvement of the final energy utilization in the residential and service sector, the diminution of energy losses in heating networks (district heating networks), as well as the substitution of conventional energy sources by unconventional ones, particularly those using renewable energy sources. This help consists of a bonus payment to the extent of 20% of the credit sum, incurred during the renovation. The bonus is paid directly to the bank providing the credit for modernization from the fund by the National Bank of Poland as repayment of the part of the credit after the execution of all renovation works.

One measure which contributed to the improvement of the energy-efficiency in the residential sector was promotion of rational energy utilization by owners of flats and single-occupier houses. Informational campaigns were initiated explaining the utility of the subject and potential profitabilities to be achieved when using energy efficient products. This led to the substitution of the lighting in residential buildings to energy-saving devices and the removal of older high energy-consuming devices in households in favour of devices which consume less energy. A further aim of this goal was the use of tax-incentives and discounts for consumers and producers of household energy efficient devices. Attention was also paid to the appropriate placement of the energy-efficiency logo on relevant products, to allow future users to identify advantages resulting from the purchase of the lower energy-consuming items.

However, the necessity still remains of conducting stimulatory and experimental research whose target is to increase energy-efficiency in the residential sector, because this sector is still an undefined energy sink for the following reasons [21]:

- The sector encompasses a wide variety of structure sizes, geometries and thermal envelope materials.
- Occupant behaviour varies widely and can impact energy consumption by as much as 100% for a given dwelling [22].
- Privacy issues limit the successful collection or distribution of energy data related to individual households.
- Detailed sub-metering of household end-uses has a prohibitive cost.

On 19 May 2010 the Directive 2002/91/EC was replaced by the Directive 2010/31/UE. The decrease of the primary energy demand and the diminution of the CO_2 emission was made possible thanks to the reduction of heat demand for heating, to the increased efficiency of energy delivery and the utilization of renewable energy. The lower energy consumption and the use of a wider range of renewable energy sources should contribute to higher safety standards in energy delivery and to support engineering development in individual countries.

In order to optimize the energy consumption in buildings, specific requirements connected: (i) with the general energy-characteristics of technical systems (installation of central heating, hot water preparation, air-conditioning and ventilation), (ii) with their appropriate installation and proper dimensioning, (iii) with regulation and inspection of these systems [23], will have to be determined.

Chapter 3

REQUIREMENTS ON NON-RENEWABLE PRIMARY ENERGY

In order to reduce the energy consumption of buildings in Poland limiting values of the indicator known as the coefficient of the primary energy demand (*EP)* [kW·h/(m^2·a)], were introduced. This indicator defines the annual computational non-renewable primary energy demand of a building for heating, ventilation, hot water preparation and for cooling. The maximum values of the indicator (*EP)* depend on the building shape factor (A/V_e) and is defined as follows [24]:

1. In residential buildings for heating, ventilation and hot water preparation:
   a) for $A/V_e \leq 0.2$

$$EP_{H+W} = 73 + \Delta EP \; [\text{kW·h/(m}^2\text{·a)}] \tag{1}$$

   b) for $0.2 \leq A/V_e \leq 1.05$

$$EP_{H+W} = 55 + 90 \cdot \frac{A}{V_e} + \Delta EP \; [\text{kW·h/(m}^2\text{·a)}] \tag{2}$$

   c) for $A/V_e \geq 1.05$

$$EP_{H+W} = 149.5 + \Delta EP \; [\text{kW·h/(m}^2\text{·a)}] \tag{3}$$

where:

ΔEP – the addition to the individual non-renewable primary energy demand for hot water generation within a year, determined by use of the Equation:

$$\Delta EP = \Delta EP_w = \frac{7800}{300 + 0.1 \cdot A_f} \text{ [kW·h/(m}^2\text{·a)]} \quad (4)$$

A_f – the useful heated surface of the building (living space), m^2,

A – sum of surfaces of all building partitions, separating the heated parts of the building from external air, the ground and adjacent unheated rooms, counted on the external contour, m^2,

V_e – net capacity of heated parts of the building, that is the gross capacity of the building diminished by the capacity of staircases, elevator, and also not closed from all sides (balconies, terraces, loggia), m^3.

In residential buildings, for heating, ventilation, cooling and hot water preparation:

$$EP_{H,C+W} = EP_{H+W} + (5 + 15 \cdot A_{w,e} / A_f)(1 - 0.2 \cdot A / V_e) \cdot A_{f,c} / A_f \text{ [kW·h/(m}^2\text{·a)]} \quad (5)$$

where:

$A_{w,e}$ – surface of external walls of the building, counted on the external contour, m^2,

$A_{f,c}$ – useful cooled surface of the building (living space), m^2.

In buildings of the group residence, in public buildings and in production buildings for heating, ventilation, cooling, hot water preparation and lighting:

$$EP_{H,C+W+L} = EP_{H+W} + (10 + 60 \cdot A_{w,e} / A_f)(1 - 0.2 \cdot A / V_e) \cdot A_{f,c} / A_f \text{ [kW·h/(m}^2\text{·a)]} \quad (6)$$

In this case for calculations of ΔEP two accessories are taken into account, i) the addition to the individual non-renewable primary energy demand on hot water preparation within a year (EP_w) (Equation 7) and ii) the addition to the individual non-renewable primary energy demand for lighting within a year (EP_L) (Equation 8).

$$EP_W = 1.56 \cdot 19.10 \cdot V_{cw} \cdot b_t / a_1 \text{ [kW·h/(m}^2\text{·a)]} \quad (7)$$

where:

V_{cw} – individual daily hot water consumption, dm^3/((j.o.)·d),
b_t – non-dimensional time of the use of hot water installation within a year,
a_1 – participation of surface (A_f) per reference unit, m^2/(j.o.).

$$EP_L = 2.7 \cdot P_N \cdot t_o / 1000 \text{ [kW·h/(m}^2\text{·a)]} \quad (8)$$

where:

P_N – reference electric power, W/m^2,
t_o – extent of lighting use, h/a.

The energy characteristics of the building, the living space or the part of the building comprising the independent technical-useful whole, not equipped with cooling installation is defined on the basis of the calculated indicator of annual non-renewable primary energy demand. The methodology for calculating the annual energy characteristics of buildings located in Poland is described in [25].

Apart from the thermal characteristics, this methodology contains other important factors, such as the kind of heating and air-conditioning system, the use of renewable energy sources, elements of passive heating and cooling, shading, the quality of indoor air, suitable natural light and the overall design of the building. A valid certificate of energy characteristics should contain information on the actual influence of the heating and cooling on the energy demand of the building, on primary energy consumption and the carbon dioxide emission.

The indicators of annual primary energy (*EP)* and the annual final energy demand (*EK)* may be determined by use of the Equations:

$$EP = \frac{Q_P}{A_f} \text{ [kW·h/(m}^2\text{·a)]} \quad (9)$$

$$EK = \frac{(Q_{KH} + Q_{KW})}{A_f} \text{ [kW·h/(m}^2\text{·a)]} \quad (10)$$

where:

Q_P – annual primary energy demand for heating, ventilation, hot water preparation and operation of auxiliary devices, kW·h/a,

A_f – useful heated surface of the building (living space), m^2,

Q_{KH} – annual final energy demand for heating and ventilation, kW·h/a,

Q_{KW} – annual final energy demand for hot water preparation, kW·h/a,

The annual primary energy demand in this case may be calculated by use of the Equation:

$$Q_p = w_H \cdot Q_{KH} + w_W \cdot Q_{KW} + w_{el}(E_{el,pom,H} + E_{el,pom,W}) \text{ [kW·h/a]} \quad (11)$$

where:

w_H – coefficient of input of the non-renewable primary energy for generation and delivery of the final energy carrier (or final energy) to the building for heating,

w_W – coefficient of input of the non-renewable primary energy for generation and delivery of the final energy carrier (or final energy) to the building for hot water preparation,

w_{el} – coefficient of input of the non-renewable primary energy for generation and delivery of the final energy carrier (or final energy) to the building - refers to electrical energy,

$E_{el,pom,H}$ – annual final electrical energy demand for the operation of auxiliary devices associated with the heating and ventilation system, kW·h/a,

$E_{el,pom,W}$ – annual final electrical energy demand for the operation of auxiliary devices for hot water preparation, kW·h/a.

The annual final energy demand for heating and ventilation can be calculated using Equation (12):

$$Q_{KH} = \frac{Q_{H,nd}}{\eta_{H,g} \cdot \eta_{H,s} \cdot \eta_{H,d} \cdot \eta_{H,e}} \text{ [kW·h/a]} \tag{12}$$

where:

$Q_{H,nd}$ – annual useful energy demand of the building (living space) by the heating system, kW·h/a,

$\eta_{H,g}$ – average seasonal efficiency of generation of the heat medium,

$\eta_{H,s}$ – average seasonal efficiency of the accumulation of heat in capacitive elements of the heating systems,

$\eta_{H,d}$ – average seasonal efficiency of the transport of the heating medium within the building,

$\eta_{H,e}$ – average seasonal efficiency of the regulation and the heat use in the building.

Partial efficiencies of the heating system could be assumed on the basis of:

- obligatory regulations,
- the technical documentation of the building, installation and devices,
- the technical knowledge and local vision of the building,
- accessible catalogue data of each element of the installation.

Table 1 shows the efficiencies of heat generation for selected heat sources [25].

Table 1. Efficiencies of heat generation for various heating systems [25]

Type of heat source	$\eta_{H,g}$
Coal boiler produced after 2000	0.82
Biomass boiler with manual service and 100 kW power	0.72
Electric direct radiators: convector, panel, radiant and floor cable	0.99
Boiler on gas or liquid fuel with open combustion space and two adjustment for regulation of the combustion process	0.86
Low-temperature gas or oil boilers with the closed combustion space and with modulated burner of power <=50 kW	0.87÷0.91
Gas condensing boiler with power <=50 kW with heating medium parameters Of 55/45°C	0.94÷1.00
Glycol/water heat pump in new/existing buildings	3.5/3.3
Compact hot-spot with enclosure and power to 100 kW	0.98
Compact hot-spot without enclosure and power to 100 kW	0.91

Efficiencies of the heat distribution as a function of the kind of heating installation are presented in Table 2.

Table 2. Average values of heat distribution efficiencies [25]

Type of heating installation	$\eta_{H,d}$
Heat source in the room (electric heating, tile stove) The housing heating (gas boiler or the mini-thermal station)	1.00
Central water heating installation supplied from the local heat source situated in heated building, with insulated pipes, armature and devices, installed in heated rooms	0.96÷0.98
Central water heating installation supplied from the local heat source situated in heated building, with insulated pipes, armature and devices, installed in unheated rooms	0.92÷0.95
Air heating	0.95

Average values of regulation and heat utilization efficiencies in the building may be assumed in compliance with Table 3.

Table 3. Average values of regulation and the heat utilisation efficiencies [25]

Type of heating installation	$\eta_{H,e}$
Direct electric heating (convector, panel, radiant)	0.98
Electric accumulation radiators (convector, floor cable)	0.90
Water heating with sectional or panel radiators in case of the adaptive and local central regulation	0.98÷0.99
Water heating with sectional or panel radiators in the case of local regulation	0.86÷0.91
Water heating with sectional or panel radiators in case of central and local regulations (range P-2K)	0.93
The floor or wall heating in case of central and local regulation	0.97÷0.98
Local heating, lack of the automatic regulation in the room	0.80÷0.85

The annual (seasonal) energy demand for heating the building ($Q_{H,nd}$) can be defined on the basis of the standard *EN ISO 13790:2009* by the Equation:

$$Q_{H,nd} = Q_{H,ht} - \eta_{H,gn} \cdot Q_{H,gn} \text{ [kW·h/a]} \tag{13}$$

where:

$Q_{H,ht}$ – heat losses by heat penetration and ventilation within a heating season, kW·h/a,

$\eta_{H,gn}$ – coefficient of efficiency of use of heat gain, estimated via Eq. (22) or Eq. (23),

$Q_{H,gn}$ – total heat gains (from internal sources and from insolation) during the heating season, from Eq. (19), kW·h/a.

Heat losses by penetration and ventilation in the heating season for a building or a flat can be determined using the Equation:

$$Q_{H,ht} = (H_{Tr} + H_{Ve})(\theta_{\text{int},H} - \theta_e) \cdot t_s \cdot 10^{-3} \text{ [kW·h/a]} \tag{14}$$

where:

H_{Tr} – coefficient of heat losses by penetration through all external partitions (the total coefficient of heat transfer by penetration), W/K,

H_{Ve} – coefficient of heat losses by ventilation (the total coefficient of heat transfer by ventilation), W/K,

$\theta_{int,H}$ – projected indoor temperature (setting of the temperature) for the heating period in the building or flat, °C,

θ_e – average outdoor temperature for the period analysed according to data from the nearest weather station, °C,

t_s – number of hours in heating season in the year, h/a.

The coefficient of heat losses by ventilation can be calculated using the Equation:

$$H_{Ve} = \rho_a \cdot c_a \sum_k (b_{ve,k} \cdot \dot{V}_{ve,k,mn}) \text{ [W/K]} \tag{15}$$

where:

$\rho_a \cdot c_a$ – volumetric heat capacity of air, which may be assumed to be equal to 1200 J/(m^3·K),

$b_{ve,k}$ – corrective coefficient for the air stream (k),

k – identifier of the air stream (air infiltration, natural ventilation, mechanical ventilation should be taken into account depending on the case),

$\dot{V}_{ve,k,mn}$ – average air stream (k), m^3/s.

For example for a building with natural ventilation can be assumed:

$$b_{ve,1} = 1;\ \dot{V}_{ve,1,mn} = \dot{V}_{o};\ b_{ve,2} = 1,\ \dot{V}_{ve,2,mn} = \dot{V}_{inf}, \tag{16}$$

and for a building with mechanical supply-exhaust ventilation:

$$b_{ve,1} = 1 - \eta_{oc};\ \dot{V}_{ve,1,mn} = \dot{V}_{f};\ b_{ve,2} = 1,\ \dot{V}_{ve,2,mn} = \dot{V}_{x} \tag{17}$$

where:

$\dot{V}_{o}$ – computational ventilation air stream as required from hygienic requirements, determined on the basis of the standard *PN-83/B-03430/Az3:2000*, m^3/s,

$\dot{V}_{inf}$ – infiltrating air stream from leaks, caused by wind and thermal buoyancy, m^3/s,

$\dot{V}_{f}$ - the higher of the two air streams: supply ($\dot{V}_{su}$) and exhaust ($\dot{V}_{ex}$), m^3/s,

η_{oc} – efficiency of heat recovery from exhaust air,

$\dot{V}_{x}$ – the additional infiltrating air stream caused by leaks by working ventilators, caused by wind and the thermal buoyancy, calculated from the dependence:

$$\dot{V}_{x} = \frac{e \cdot n_{50} \cdot V}{1 + \frac{f}{e} \cdot \left[\frac{(\dot{V}_{su} - \dot{V}_{ex})}{n_{50} \cdot V} \right]^2} \cdot \frac{1}{3600}\ [m^3/s] \tag{18}$$

where:

e, f – the coefficient of building protection (Table 4),
n_{50} – air change rate in building caused by pressure difference 50 Pa, h^{-1},
V – indoor ventilated capacity, m^3,

The kind of the ventilation has a very strong influence on the primary energy demand, on the air infiltration by the building structure and on the energy quality of the building. The contribution of thermal needs on the ventilation in the energy balance of the building amounts to 30÷60% [26-28].

Table 4. Coefficients of building protection (*e*), (*f*) [25]

Coefficients (*e*) for protection class:	**More than one unprotected façade**	**One unprotected façade**
- unprotected: buildings on an open space, high buildings in centres of cities	0.10	0.03
- protected: buildings among trees or other buildings, buildings in suburbs	0.07	0.02
- highly protected: medium-high buildings in cities, buildings in forests	0.04	0.01
Coefficients (*f*)	15	20

In the energy balance of the building or the flat one should take into account internal heat gains (Q_{int}) and insolation (Q_{sol}) during the considered time period, in accordance with equations (Eqs 19÷21):

$$Q_{H,gn} = Q_{\text{int}} + Q_{sol} \text{ [kW·h/a]} \quad (19)$$

$$Q_{\text{int}} = q_{\text{int}} \cdot A_f \cdot t_s \cdot 10^{-3} \text{ [kW·h/a]} \quad (20)$$

where:

q_{int} – thermal load of rooms with internal gains (for existing buildings the average of the individual power of internal heat gains can be assumed in compliance with the Table 5), W/m^2,

A_f – surface of rooms with regulated temperature in the building or the flat, m^2.

Table 5. Approximate thermal power of internal heat sources referred to the surface (A_f) [25]

Kind of building	**q_{int}, W/m^2**
One-family building	2.5 ÷ 3.5
Residential building (flat)	3.2 ÷ 6.0
Schools	1.5 ÷ 4.7
Offices	3.5 ÷ 6.4

Heat gains from insolation should be balanced for each month separately, and then summed for the heating season. Monthly heat gains from the Sun through glazed surfaces installed in vertical partitions and in the roof, are calculated by use of the Equations:

$$Q_{sol} = \sum_i C_i \cdot A_i \cdot I_i \cdot g \cdot k_\alpha \cdot Z \text{ [kW·h/m-c]} \quad (21)$$

where:

C_i – the participation of glazed surface in entire surface of window, depends on sizes and construction of the window, the mean value amounts 0.7,

A_i – glazed surface in the light of opening in partition, m^2,

I_i – value of solar radiation energy in the month of the heating season being considered on the vertical plane, in which the window is situated, according to data for the nearest actinometric station, kW·h/(m^2·m-c),

g – coefficient of permeability of solar radiation energy through the glass, in Table 8,

k_α – corrective coefficient taking into account the angle of inclination of the roof surface in relation to the level (for the vertical partition $k_\alpha = 1$, in other cases, this value should be accepted according to [25]),

Z – coefficient of shading of the building, according to Table 6.

Table 6. Values of coefficient of shading of various building types [25]

Location of the flat	***Z***
Buildings on the open space or high buildings in the centres of cities	1.00
Flats in buildings like upper, in which at least the half the windows are shaded by elements of the building (loggia, balcony of the neighbour)	0.96
Buildings in cities close to buildings of the same height	0.95
Triple pane glazing in short and medium buildings in the centres of cities	0.90

The use of the coefficient of efficiency of use of total heat gains ($\eta_{H,gn}$) in the heating mode is estimated using the Equations:

for: $$\gamma_H = \frac{Q_{H,gn}}{Q_{H,ht}} \neq 1$$

$$\eta_{H,gn} = \frac{1-\gamma_H^{a_H}}{1-\gamma_H^{a_H+1}} \tag{22}$$

- and for: $\gamma_H = 1$

$$\eta_{H,gn} = \frac{a_H}{a_H + 1} \tag{23}$$

where:

a_H – a numerical parameter, depending on the time constant of the building or its zone, may be calculated:

$$a_H = a_{H,0} + \frac{\tau}{\tau_{H,0}} \tag{24}$$

where:

$a_{H,0}$ – dimensionless reference coefficient equal to 1.0,
τ – time constant for zone of the building or the total building, h,
$\tau_{H,0}$ – reference time constant equal to 15 h.

As a result of research under Finnish weather conditions, other values of numerical parameters, solely for use with residential buildings, were estimated and are equal to $a_{H,0}$=6 and $\tau_{H,0}$=7 h [29].

The time constant of the building or its parts can be defined with the Equation (25) depending on the internal heat capacity of the building (C_m).

$$\tau = \frac{C_m}{H_{tr,adj} + H_{ve,adj}} \cdot \frac{1}{3600} \text{ [h]} \tag{25}$$

$$C_m = \sum_j \sum_i (c_{ij} \cdot \rho_{ij} \cdot d_{ij} \cdot A_j) \text{ [J/K]} \tag{26}$$

where:

c_{ij} – specific heat of material of the *i-th* layer in the *j*-th element, J/(kg·K),
ρ_{ij} – density of material of the *i-th* layer in the *j*-th element, kg/m^3,

d_{ij} – thickness of the *i*-layer in the *j*-th element, where the total thickness of layers can not exceed 0.1 m,

A_j – area of the *j-th* element of the building, m^2.

The time constant of the building characterizes the ability of the building construction to accumulate heat and the compensation of temperature oscillations in heated rooms.

Norén et al. [30], on the basis of simulations, ascertained that it is possible to decrease thermal needs for heating by about 16÷18%, if the building has a large thermal inertia (τ = 325 h) in comparison to a building with a low thermal inertia (τ = 31 h). In [31] it was shown that a building's thermal inertia has a significant impact on its energy consumption.

Multi-criterion optimization of residential buildings in terms of the rational heat demand for heating as a function of their shape, was introduced in the study by Jedrzejuk and Marks [32-34].

Depecker et al. [35] and Catalina et al. [31] on the basis of simulations defined the dependence between building shape factor and energy consumption.

Jaber and Ajib [36] assessed the best orientation of the building, windows size, thermal insulation thickness from the energetic, economic and environmental points of view for a typical residential building located in the Mediterranean region. The results showed that about 27.6% of annual energy consumption could be saved by choosing the best orientation, optimal window size and shading device, and optimal insulation thickness.

Chapter 4

ENERGY EFFICIENT SOLUTIONS

Bearing in mind that both increasing the energy-efficiency in the residential sector as well as comprehensive public information on energy efficiency and energy- and system-related costs of various heat supply systems are crucial [37], in this chapter we describe energy efficiency systems, applications and solutions. This information will be of particular use to practitioners, because, applying it, they will be free to choose the best system (the cheapest and most efficient one), for their special needs [38]. Current regulations concerning thermal protection for newly designed residential buildings are defined in technical requirements, which buildings and their location should fulfil [24].

4.1. BUILDING ENVELOPE

Installation of a suitable thermal insulation for building elements is the simplest way to limit heat losses by heat-transfer. Heat transfer through a building involves heat conduction and absorption of heat on both surfaces of a building partition. In compliance with the standard *EN 6946:2008,* the coefficient of heat transfer of building partitions can be estimated by use of Equation (27).

$$U = \frac{1}{R_{si} + \sum_i R_{\lambda i} + R_{se}} \ [\mathrm{W/(m^2 \cdot K)}] \qquad (27)$$

where:

R_{si} – resistance of heat transfer on the internal partition surface, (m^2·K)/W,

$R_{\lambda i}$ – thermal conduction resistance of the *i*-th layer building partition, (m^2·K)/W,

R_{se} – resistance of heat transfer on the external partition surface, (m^2·K)/W.

Parameters of external partitions in new and renovated buildings have been continuously made more stringent for the purpose of the diminution of the thermal energy consumption. This is evident from recent changes in the Polish legislation, whose use is obligatory in the design and calculation of buildings (Table 7).

Table 7. Changes in the required U_{max} of buildings in Poland: before and after the change in regulations on 06. November 2008

Type of partition	Maximum value of heat-transfer coefficient U_{max} [W/(m^2·K)]						
	PN-57	PN-64	PN-74	PN-82	PN-91	Before change	After change
External wall	1.16	1.16	1.16	0.75	0.55÷0.70	0.30÷0.50	0.30
Roofs, flat roofs	0.87	0.87	0.70	0.45	0.30	0.30	0.25
External window	-	-	-	2.0÷2.6*	2.0÷2.6*	2.0÷2.6*	1.7÷1.8*

* the higher value for I, II and III the climatic zone; the lower value for IV and V the climatic zone

The floor in a heated room should have a circuital heat insulation layer. These layers should have a minimum thermal resistance of 2.0 (m^2·K)/W.

Implementing these improvements causes a considerable diminution of the design heating load and average energy consumption per m^2 of flat surface in new buildings. In Poland this amounts to even twice that of the average heat consumption of existing buildings. The most frequently applied insulating materials in Poland are polystyrene foam and rock wool [39].

An important problem is selecting the appropriate thickness of the isolation for building partitions in newly designed or renovated objects. Generally, an investor will attempt to achieve a balance between costs connected with the implementation of this thermal solution and the profits accruing due to the diminution of heating costs. In the study [40], it was noted that the lower the thermal resistance of a non-insulated partition, the higher the profitability of the renovated partitions. In addition, for partitions with lower thermal resistance the amortization time of the investment cost is shorter. From an economic standpoint, the thickness of the isolation of external partitions should be enlarged to 0.2 m, although nowadays in most cases a thickness of 0.1 m is applied [41,42].

In the study [43], a new decision support system for the integrated assessment of thermal insulation solutions with the emphasis on recycling potential is presented. This solution comprises three main assessment factors: primary energy consumption, the environmental impact and the financial cost.

Also, in the study [44] the authors searched for an optimized insulation thickness by minimizing the total cost of insulation and energy consumption. They showed that the optimum thickness of a single insulation layer is independent of its location in the wall. The best overall performance is achieved by a wall with three layers of insulation, each 26-mm-thick, placed at inside, middle and outside followed closely by a wall with two insulation layers, each 39-mm-thick, placed in the middle and outside.

In the case of already existing buildings the diminution of the energy consumption used for heating, could be obtained by undertaking thermal renovation, including improvement of the building structure. Thermal renovation is essential, first of all because of the rise in prices of energy carriers and their decreasing availability, as well as of in the interest of environment protection and human health.

Carefully executed thermal insulating (new objects) and complex thermal renovation (existing objects), lead to heating costs lowered by about 50%, and a simple payback time of 7 years [45]. Thanks to thermal insulating, householders can achieve a considerable reduction in the heat-transfer losses through building partitions. This results in a situation in which the fraction of heat used to heat ventilating air makes up about 20÷50% of the total heat demand for the building.

The problem of thermal bridges should not be forgotten. These should be removed at the early design stage and then carefully supervised for correct implementation later in the execution stage. Thermal bridges are caused across the discreteness or the diminution of the thickness of the thermal insulation and the irregular execution of external partitions of the building. They can lead to condensation of water vapor and the development of mould on the interior surface of building partitions, as well as to substantial heat losses to the environment. Detection of thermal bridges, leaks, humidity can be executed without invasion by means of thermography [46]. Thermographic inspection, consisting of the visualisation of the temperature field of internal and external building partitions, permits construction firms to undertake decisions which may lead to improvements to the heat-insulation of the building: These, in turn, can lead to the diminution of heating costs and improvements in thermal comfort [47]. The infrared thermography method which is used to visualize the thermal bridges as well as a complementary experimental method allowing for the determination of the quantitative aspects of the heat losses through the envelope, was discussed in the study [48].

Besides external walls, the most important element of the building which influences heat losses to the environment, is the window. Windows play an important role in the energy performance and, therefore, should be chosen carefully. Defining a Window Energy Rating System (WERS) is not a simple matter because window performance depends on the climate in which the window will be used, the type of building and the orientation. There are many different rating systems and methods [49-51], which are being applied in several countries. For example the English model [52] uses these properties for the window, combined into a single equation. The National Fenestration Rating Council (NFRC) system [53], on the other hand, considers the window properties, thermal transmittance (U-value), solar heat gain coefficient (SHGC), visible light transmittance (VT) and air leakage (AL).

The Danish model [54] uses the U-value and g-value of the glazing, combining them into a single equation obtained from the energy balance, considering a distribution of windows in a single family house. In addition to this indicator, this model also uses the U-value of the frame, multiplied by the design width of the frame, and the linear thermal transmittance of the glazing edge in order to rate a window according to energy performance.

In [55] it is proved that double-glazing with a low emissivity coating and high solar heat gain is very appropriate in temperate and cold climates, and that the influence of the glazing area is more significant in cold climates than in temperate ones.

Industries have developed window systems that can be completely reversed through their sashes to reduce glass-cleaning costs and to provide emergency escapes, especially in upper storeys and highrise buildings. Nowadays, reversible windows are popular also for applications in residential buildings and small office spaces, as the glass rotation takes place outside the frame, leaving curtains, blinds and ornaments undisturbed, without intrusion into the room.

Windows currently on the market are normally only reversible to a partly open position for cleaning purposes, and will have to be further developed in order to function as closed windows in the reversed position. Double-glazed windows with a clear and an absorbing glass pane are commercially available with an absorbing pane which is able to reject a large portion of the absorbed solar radiation to the external or to the internal environment, depending on whether it faces out- or indoors.

In the study [56] it was shown that double-glazed window systems made of an absorbing and a clear glass pane can reduce yearly energy requirements if they are turned by 180°, leaving the absorbing pane facing the indoor side during the warmer part of the year, and the outdoor side during the cooler seasons. In this case, a large fraction of the solar radiation absorbed with the colored pane can be gained in winter and rejected in summer.

The yearly energy savings obtained with reversible windows are strictly linked to the method used to control indoor air temperature to avoid intolerable overheating. It was also concluded that reversible windows facing West and East increase the yearly energy savings from 10% to 15% in respect to the previous results obtained for South orientation. This is mainly due to the lower amount of energy wasted to control winter overheating.

In the study [57] the impact of different kinds of glazing systems, window size, orientation of the main windowed facade and internal gains on winter and summer energy need and peak loads of a well insulated residential building, were evaluated. The authors concluded that in winter the use of windows with low thermal transmittance is useful if accompanied by high solar transmittance. Also the selective shading systems should be installed to improve summer performance without affecting the winter one.

The frame position and the configuration of the window hole insulation have a significant impact on the overall thermal performance of the window [58].

In Poland the windows area, glass- and transparent partition with a coefficient of heat transfer of at least 1.5 $W/(m^2{\cdot}K)$, in both residential buildings and group residences should not exceed the calculated value [24]:

$$A_{O\max} = 0.15 \cdot A_z + 0.03 \cdot A_w \ [\mathrm{m}^2] \tag{28}$$

where:

A_z – sum of or floor projection surfaces of all storeys (not including a cellar) at a width of 5 meters in the external outline of the building, m^2,

A_w – sum of surfaces of the remaining part of the horizontal projection after deduction of A_z, m^2.

According to the French Thermal Standard FTS the optimal value of the window to floor area ratio is 16.5% and should not exceed 22% [31].

In all kinds of buildings, the transmittance coefficient of total energy for windows and glass- and transparent partitions (g_c), calculated using Equation (29), should not be greater than 0.5.

$$g_c = f_c \cdot g_G \tag{29}$$

where:

f_c – the corrective reduction coefficient of radiation in consideration of applied sun-devices,

g_G – transmittance coefficient of total energy for the type of glazing, defined as in Table 8.

Table 8. Dependence of the total energy transmittance coefficient (g_G) on glazing type [24]

L.p.	Glazing type	Coefficient (g_G)
1.	Single glazing	0.85
2.	Double glazing	0.75
3.	Double glazing with selective low-emissivity coating	0.67
4.	Triple glazing	0.70
5.	Triple glazing with selective low-emissivity coating	0.50

Whenever the participation of glazed and transparent surfaces (f_G) in the wall surface exceeds 50%, the following dependence should be fulfilled:

$$g_c \cdot f_G \leq 0.25 \tag{30}$$

Among situations under which the values of transmittance coefficient of total energy for windows and glass- and transparent partitions (g_c) can be neglected are the following:

- vertical surfaces and inclined surfaces more than 60° in relation to the level, in the northern orientation ± 45°,
- windows protected from the solar radiation by the artificial partition or the natural building partition,
- windows with surfaces smaller than 0.5 m^2.

The thermal properties of the building are closely connected with its resistance to air penetration. This resistance consists of the ventilation of the building by specially placed openings in external partitions and the unwanted infiltration of external air through slits in the building substance [59]. Such leak-tightness is depends on the pressure differential between both sides of the given partition, on the wind direction, on the efficiency of the ventilating system and from the difference in indoor and outdoor temperatures.

Externally opaque partitions of buildings, joints between partitions and parts of partitions and connections of windows with casings should be characterized by the total tightness on the air penetration. In residential buildings, collective residences and in public buildings the coefficient of air infiltration for open windows and balcony doors should not exceed 0.3 $m^3/(h \cdot m \cdot daPa^{2/3})$. For buildings with gravitational ventilation the required tightness (n_{50}) should not be greater than 3.0 h^{-1}, and for buildings with mechanical ventilation not greater than 1.5 h^{-1} [24].

Also the external rolling shutter can be a source of energy saving, because it provides an additional barrier to heat-transfer. This solution separates the internal environment from the external one and is assembled outside the windows, which contributes to improvements in the thermal parameters of the most disadvantageous element of the partition in respect of thermal losses, namely the window. It was also assumed that external rolling shutters would be closed during night hours in winter to reduce the heating loss of the indoor space. In summer, it is always opened to take advantage of the lower night temperature and to reduce the load on the cooling system in the early morning, when the system begins to work.

The financial costs of such measures, which are often a limiting factor, ought not to be neglected when considering the reduction of energy consumption in existing buildings. For this reason, priorities have to be established to identify the most profitable activities, i.e. those which lead to the highest energy savings over the longest period of time. One hierarchy of such priorities is as follows roof isolation, isolation of the floor, use of more efficient thermal glazing by ”updating” existing windows.

4.2. Heat Supply, Distribution and Producing Systems

Presently there are many technical solutions which make the delivery of separate heat and power (SHP) to the building. Also possible is the technology known as combined heat and power (CHP). The European Union (EU), in their cogeneration strategy of 1997, set an overall target of doubling the share of electricity production from cogeneration to 18% by 2010 [60]. Also the South Korean government has placed high priority on CHP/DH support programs in its energy conservation policy. District heating (DH) operators there are given the so-called „heat supply monopoly" mostly in newly developed residential areas [37].

However, sustainable development of a local heat market requires application of planning procedures which include not only optimization of both the demand and supply side of the local energy market but also experimental results, which give information about performance and characteristic attributes of any given heating system.

The methods of local energy planning can be classified in three separate categories [61]:

- planning by models: can be based either on econometric or optimization models. Econometric models utilize mathematical or statistical methods and rely on statistical data. Optimization models allow for the identification of best possible solutions – minimization or maximization of an objective function, with a predefined set of constraints which describe the space of acceptable solutions.
- planning by analogy: which utilizes a simulation of the energy market in less developed countries by the behaviour of those in well developed countries. This kind of local energy planning is usually used for the verification of planning results achieved by other planning methods [62].
- planning by inquiry: is used in the case when the other methods mentioned above are not reliable. This method may be based, for example, on a questionaire answered by a group of energy market experts and the statistical evaluation of their responses.

For example in the study [63] a new approach to planning the modernization and development of community heating systems was employed for a case study analysis referring to the choice of the heating system for a new developing urban area (12 residential buildings, each consisting of 10 separate apartments). As the main finding it was noted that, from among five analysed systems, a system based on a district heating system supplied by a municipal CHP plant or island type heating system based on natural gas condensing boilers, should be taken as the recommended solution.

However, it should be noted that, in the case of buildings supplied with heat from a heat distribution network, the problem of heat losses on transportation in the section between the building and the heating plant or the heat and power generating plant, appears. These losses amount, for channel-networks, to about 14% of the transported heat and, within the summer period, to even 30% [64,65]. The efficiency of the heat insulation of pipelines in the channel-network after a dozen years of exploitation falls to even ca. 50% in comparison to the initial value [64]. Besides the piping, heat losses in the heat distribution networks also occur in heat chambers, where usually the not-isolated cutting off armatura are situated. In the study [66] it was estimated that heat losses in heat chambers can amount to about 20÷30% of the total heat losses in the heat distribution network.

In this case, to reduce generation and distribution inefficiencies, decentralized energy systems will be favoured over more traditional centralized ones [67]. This means that the heating source for the building can, in accordance with the decentralization rule, be a gas, oil, or biomass boiler, because using one of these will limit transfer heat losses from the heat and power generating plant which provides between 8 and 10% of the heat generated [68]. In the study [37] it was mentioned, that DH is not cheaper than individual gas heating by condensing boilers in terms of energy costs. DH is a very expensive option when system-related costs such as depreciation and maintenance costs, partial financing of investments in DH facilities by users and various governmental supports, are considered. Also, DH is not environmentally friendlier than individual gas heating by condensing boilers.

There are also arguments in favour of supplying this system from the district heating network, which, in Wojdyga's opinion [69], is cheaper and more ecologically friendly than establishing local boiler rooms because large sources producing heat for heating systems are generally equipped with high efficiency units thus reducing the emission of combustion products into the atmosphere.

In Poland minimum-requirements concerning thermal isolation thicknesses of distributive pipes and elements of the central heating installation, the hot water installation, and cooling installation and air heating are presented in the Table 9, in compliance with [24].

In the case of complex thermal renovation it should be remembered that, in addition to further thermal insulation of external partitions and changing the windows, complete modernization of the central heating installation will be essential. Unfortunately, often when building or retrofitting a house, the heating installation is considered last. People do not really seem to care too much about the performance of the installed system as long as it provides the desired indoor temperature [70,71].

Table 9. Requirements concerning the thermal insulation of pipes and chosen elements of installation [24]

L.p.	Type of pipes or elements	Minimum-thickness of thermal insulation λ = 0.035 W/(m·K)[1]
1.	internal diameter to 22 mm	20 mm
2.	internal diameter from 22 to 35 mm	30 mm
3.	internal diameter from 35 to 100 mm	equal to internal pipe diameter
4.	internal diameter more than 100 mm	100 mm
5.	pipes and the armature according to the item 1-4 passing through walls, ceilings, or crossing of pipes	½ requirements from item 1-4
6.	air heating ducting (arranged inside the heat insulation of the building)	40 mm
7.	air heating ducting (arranged outside the heat insulation of the building)	80 mm
8.	pipes of ice-water installation inside the building [2]	50% requirements from item 1-4
9.	pipes of ice-water installation outside the building [2]	100% requirements from item 1-4

1) by use of the insulating material with a heat conduction coefficient other than the one given in the table should properly correct the thickness of thermal insulation layer,
2) thermal insulation made as airproof.

Among other things particularly important is the heat source, which will be adapted to the current thermal needs of the object. In the study [72], it was shown that biomass boilers bring the highest economic impact. However, they are dependent on fuel accessibility and they demand frequent supervision. The coal-fired boiler is however least ecological, and the process management of the combustion causes most problems, similarly to the biomass solid fuel boiler. Traditional solid fuel boilers used in older central heating systems, are characterized by low heat efficiencies in the range of 0.50÷0.75 and with high emission rates of gaseous and dust pollutants, which are contained in combustion gas. However, boilers fitted with upper combustion technology are highly efficient, with a furnace energy efficiency of about 90%.

Volatile products of degassing emerging from the high-temperature zone of the heat source, are almost completely combusted, resulting in low pollutant emission. The retort boiler is the most effective item in coal-boilers. They are equipped with a retort furnace, to which a portion of the fuel is delivered and with the distribution system for primary and secondary air. Delivery of a suitable quantity of solid fuel to the boiler takes place by means of a cochlear feeder impromptu continuous and automated. The electronic driver of the boiler controls the work of i) the circulation pump of the central heating, ii) the loading pump in the installation of the hot water preparation and iii) of the feeder and the ventilator. The driver makes possible the temperature stabilization of heated water and controls the fuel combustion process in the boiler, not allowing it to extinguish.

Another alternative is the gas boiler, whose exploitation is simpler thanks to automatic control of the combustion process. Further, its price is lower than that of biomass boilers.

In the study [73] research on a commercial balanced-flue space heater with a gas burner allowed the following conclusions:

- temperatures up to 600°C on the hot surface that transfers heat to the room were measured, the high values being a consequence of poor radiation and convection transfer;
- the combustion chamber external surface presents an average emissivity as low as 0.18 (silver colour), reducing thermal radiation significantly;
- closed furnace cabinets (preventing contact with hot surfaces) substantially reduced the heat transfer to the room;
- chimney airflows ranging between two and eight times higher than those necessary for complete gas combustion were found, producing very high heat losses.

Continuing the research [74] on this type of space heater, the authors made changes in heat transfer design features and implemented passive chimney control in the framework of simple and low cost modifications. The authors have not introduced any modifications to the gas supply and/or security systems as, in order to do this, licenses must be obtained from the gas authorities.

As a result they achieved a significant increase in efficiency by simple modifications, because the improvements led to better radiation and convection transfer, and with the further introduction of an innovative passive chimney-flow controller, the thermal efficiencies obtained were close to 85% in comparison with the original efficiencies, which were in the range of 39÷63%.

An improved effect for gas fuel can be achieved by using a condensing gas boiler, whose construction is still under development. For example, the authors [75] investigated experimentally a cylindrical multi-hole premixed burner for its potential use as a condensing gas boiler.

Condensing boilers are a kind of low-temperature heating device in which, by cooling the combustion gas, the heat of condensation of the water vapor contained in the condensate, is used. The efficiency of a condensing boiler is highest, when the temperature of water in the boiler does not exceed 50°C. Simultaneously, the temperature of combustion gas exhausted from boilers is low (ca. 40÷80°C). In contrast to conventional boilers, here there are no limitations regarding the minimum-temperature of water returning to the boiler. The lower it is, the better the utilization of the condensation of the water vapour and the higher the efficiency of the boiler. Thanks to the condensations of the water vapour contained in the combustion gas, heat recoveries even > 11% or more in comparison to the heat efficiency obtained in traditional gas boilers, can be achieved.

The boiler efficiency may vary considerably with load, but it is possible to set up the boiler optimally, as described in detail in the study [76].

To meet the energy and environmental requirements condensing boilers with premixed burner have recently been introduced [77,78].

However, electrical energy is not a very profitable heat source for heating systems in consideration of costs [79] and the load of the urban power grid. The use of this form of energy as a heat source for heating is legitimate in buildings, which are characterized by low heat demand, where the use of other heat sources would tie in with the low efficiency.

In the study [80] 17 different technological options of water-heating, space-heating and cooking were analyzed. The different options analyzed included a natural gas central water- and space-heating system, a natural gas cooker, a natural gas water heater, an electric heat pump, an independent electric heater, an independent electric storage heater, an electric water heater and an electric cooker.

The results showed that the use of electric heat pumps, both for space and water-heating, combined with the use of a natural gas cooker, leads to the lowest energy consumption and to the lowest environmental impact in terms of carbon emissions. Considering only the running costs, this choice is 45% more economic than having a natural gas centralized heating system combined with a gas cooker, and is 60% more economic than having an electric resistance space heater combined with an electric storage water heater and electric cooker, which is the worst case. On the other hand, the variation of the life cycle cost, as a function of the equipment cost and the energy operating costs, showed that the combination of a natural gas water heater, with an independent electric storage heater, and a natural gas cooker has the best economical performance.

In other study [81] the authors investigated the life cycle impacts of three residential heating and cooling systems: warm-air furnace and air-conditioner, hot water boiler and air conditioner, and air–air heat pump over a 35-year period. They concluded that the boiler and air conditioning system have the largest impacts associated with the appliances and distribution systems. The heat pump has its maximum impact in regions where a high proportion of the electricity is derived from fossil fuels. This could be minimized if 15÷40% of the current electric grid generation capacity were substituted by renewable energy sources.

From the environmental point of view micro-CHP is a reasonable alternative to traditional heating systems [82]. For this reason, the choice of a heating system is often determined based on its life cycle primary energy analysis [83,84].

Besides the heat source of the building or flat, attention has to be paid to the pipes which distribute the heating medium in the building, especially to those, which are continued outside the heated flats. These pipes must be properly insulated, as otherwise substantial heat losses along their length will result.

When replacing heating systems in older multi-storey houses, it is possible to achieve a high level of insulation. In connection with single-family housing, non-insulated district heating units are an area with a large potential for energy savings.

To increase the energy efficiency of the heating system, one can remove or reduce the number of distributing pipes in multi-family buildings by means of decentralization. In this case we are predisposed, among other things, to use individual bi-functional gas boilers for each single flat. The hot water and the heating medium supplying the central heating installation are then set the required thermal parameters in this device. Transportation losses are thus avoided. Another way towards the decentralization of heat supply of flats in multi-family buildings is the use of residential thermal stations, which possess a number of advantages [85], for example:

- Reduction of possible contamination of the source of hot water production by the Legionella bacterium, since the hot water in the residential thermal station (RTS) is produced from fresh cold water using an indirect heat exchanger;
- Limiting heat losses in comparison with hot water storage vessels, because the standby loss is eliminated (improvements of about 31% [86]);

- Reduction in the number of pipes (from five to three) and the number of heating risers (from an average of about a dozen to only one installed on the staircase), because a horizontal heating system is installed in every flat [87]. This solution removes the need for a hot water circulation pump thus reducing the consumption of electrical energy and eliminating this additional cost;
- Metering the consumption of thermal energy in every flat, which encourages energy savings by the occupants;
- The possibility of using the central heating installation in those seasons when heat is not usually supplied; something which is generally impossible with most other systems.

The residential thermal station (RTS) is a device in which the running water, by ensuring its priority in relation to the central heating installation, is heated locally in the plate heat exchanger. The RTS is supplied with heat from the local boiler of the building or from the mono-functional district heating station.

In the study [88] the authors present results of experimental research on three commonly used heating systems in multifamily buildings. A central heating and central domestic hot water system (system A), a decentralised system using residential thermal stations (system B) and a decentralised system using bi-functional gas boilers (system C) were studied. The average annual efficiencies for systems A, B and C were 59.6%, 70.1% and 90.5%, respectively.

4.3. Application of Heat Pumps

The heat pump becomes a key component in an energy recovery system with great potential for energy saving, especially in the present situation, when the cost of energy continues to rise. About 80% of the units installed worldwide are used in domestic appliances [89]. These are devices making possible the utilization of practically unlimited resources of renewable and waste energy. The superheat and the heat of condensation of steam is used to heat water or air in heating installations in the building.

The functioning of heat pumps is based on the lifting of the thermal potential which consists in receiving the heat from a source with a lower temperature (T_{down}) (bottom source) and delivering it to the source with the higher temperature (T_{top}) (upper source). In order for this process to run, the system must be supplied a contribution of the driving energy, which should be possibly low, in consideration of the heat pump efficiency.

Applied heat pumps can be classified as follows [90]:

- thermoelectric heat pumps,
- absorptive heat pumps,
- compressor heat pumps.

Each of the above types of heat pumps is currently receiving considerable attention from researchers all over the world and hence continuous technological progress can be expected [for up-to-date reviews see 91,92].

Most widely accepted are compressor heat pumps which, despite their known defects, have been established for many years. In this type of heat pump the heat is transported by a heating medium (water, brine), which circulates in a closed circuit based on Linde's circuit [93], presented in Figure 4.

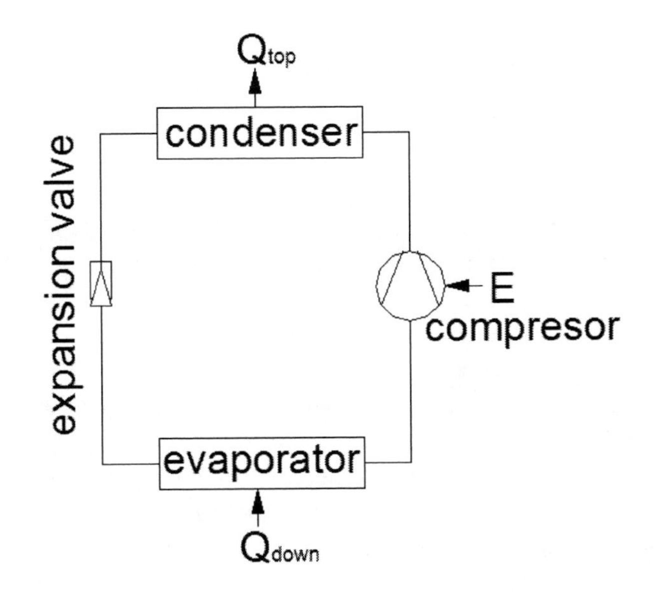

Figure 4. Scheme of compressor heat pump.

The mechanism of action of the compressor heat pump is as follows; the liquid working factor under low pressure is led to the evaporator. In the evaporator the temperature is higher than the boiling temperature of the working factor and as a result is vapourized at the cost of heat received from the low-temperature source. Subsequently, steams of the working factor are sucked in by the compressor which compresses it and, as a result, the pressure and the temperature of the steam rise. The working factor is then sent to the condenser, where it returns heat to the heating medium, in consequence of which the working factor chills and condenses. After this the liquid working factor under high pressure flows through the expansion valve, where it expands to the low pressure and its temperature drops. In this way the circuit is closed.

The coefficient of heat pump performance is the relation of the heat power of the device, which is transferred to the installation, to the power received from the power grid, the gas network or other source. It depends, among other things, on the temperature difference between the heating medium, and the bottom low-temperature heat source. The greater the difference, the lower is the efficiency. *COP* is not constant, because the temperature of bottom low-temperature heat sources in oscillates the smaller or in the greater range. The coefficient of heat pump performance can be calculated by use of Equation (31).

$$COP = \frac{Q_{top}}{E} \tag{31}$$

where:

Q_{top}- thermal power transferred by the heat pump in the upper source, kWh,
E – electrical energy supplied to heat pump, kWh.

The coefficient of heat pump performance does not take into account the use of an additional heat source, which can be, for example, the gas boiler supporting the heat pump by the top thermal demand or the energy used by circulating pumps installed in heating installation and in bottom low-temperature heat source. Therefore, by defining the overall costs spent on the heating using a heat pump the degree of installation efficiency may be applied, which is assigned the relation between the heat power transferred in the upper source by the heat pump, to the power used by the heat pump from other sources and all other devices applied in the heating installation.

Approximate values of real coefficient of heat pump performance are equal:

- for the convective heating with sectional radiators at the temperature on supply t_{supply}= 60°C: φ_r= 2.5÷3.5;
- for radiation heating by t_{supply}= 50°C: φ_r= 3.5÷4.5;

- for the hot water installation by hot water temperature t_{HW}=45°C: φ_r= 3.5÷4.0;
- for heating of basinal water:
    - in summer: φ_r= 6.0÷8.0;
    - in winter: φ_r= 4.0÷6.0.

The *COP* values for various heat pump types amount to [94]:

- air-to-air heat pumps: *COP* = 1.5÷3.9;
- water to air heat pumps: *COP* = 2.0÷4.0;
- water-to-water heat pumps *COP* = 3.0÷5.0.

The *COP* value of the vapour compression heat pumps is much higher than thermally driven pumps. This is the major advantage of the mechanically driven heat pumps over thermally driven heat pumps, for which the working principle, problem and solutions are discussed in the study [95].

The factors that can affect the life-cycle efficiency of a heat pump, are:

- local method of electricity generation,
- type of heat pump,
- size of the heat pump,
- refrigerant used,
- thermostat controls,
- quality of work during installation,
- energy efficiency of the home's layout, insulation and ducts,
- climate.

In order to improve the heat pump efficiency a multistage system can be applied, which employs more than one compression stage. Multistage vapour compression systems can be classified as compound or cascade systems [96]. A multistage system has a smaller compression ratio and higher compression efficiency for each stage of compression, greater refrigeration effect, lower discharge temperature at the high-stage compressor, and greater flexibility in comparison to single-stage systems [97,98].

The authors [99] found that a double-stage coupled heat pumps heating system, in which an air source heat pump was coupled to a water source heat pump, improved the process energy efficiency ratio by 20% in comparison to a pure air source heat pump.

Another possibility to minimize the energy consumption is to improve the performance of the compressor. That is why, for example the scroll compressor is more efficient than the standard reciprocating compressor. This was demonstrated in the case of heat pumps by [100,101]. The innovative design [102,103], used in a revolving vane (RV) compressor, involved the radical use of a rotating cylinder that moved together with the compressing mechanism to cut down on energy loss, – this reduced the energy input by as much as 80% when compared to current systems on the market.

The use of a heat pump driven by the gas engine also reduces the electricity consumption in the cooling and heating seasons and makes possible the easy modulation of the compressor speed by adjusting the gas supply [104,105]. Results presented in the study [106] show that the gas engine heat pump can save more energy than the electricity engine heat pump. In summer, the maximum percentage saving of primary energy for a gas heat pump is 60.5%. The authors concluded also, that the percentage energy saving in a frost area is higher than in other areas.

Heat pumps can also be divided depending on the nature of the low-temperature factor and the factor heated in the condenser:

- water - water (W-W),
- water – air (W-A),
- air - water (A-W),
- air - air (A-A).

Under the term „water" we understand any liquid heat factor, for example in the form of the mixture water-glycol. Similarly in the case of „air", one should understand also exhaust gases, combustion gas and other kinds of gases which can supply warmth to the evaporator of the heat pump.

The dimensioning of the heat pump consists in the definition of the thermal energy, which should be delivered by the heat source. The thermal power consists of the projected thermal load for a given building and the additional heat for hot water preparation. The heat pump is, for single family dwellings, normally sized to cover 50÷70% of the maximum designed heat load of the building. This design means that the heat pump covers 85÷95% of the annual heat demand in the building [107].

The projected thermal load for the given building should be calculated accurately according to *EN 12831:2006* [108]. However the quantity of the necessary energy for hot water preparation can be calculated using Equation (32).

$$E_{HW} = c_{HW} \cdot \rho \cdot Q_{HW} \cdot (t_{HW} - t_{CW}) \text{ [kJ/d]} \quad (32)$$

where:

E_{HW}- energy necessary for hot water preparation, kJ/d,
c_{HW}- specific heat of water; may be assumed c_{HW}= 4.2 kJ/(kg·K),
ρ- density of water; may be assumed ρ = 1000 kg/m^3,
Q_{HW}- projected hot water consumption, m^3/d,
t_{HW}- hot water temperature on outflow from the heater, °C,
t_{CW}- cold water temperature, °C.

The hot water temperature should be taken from the range 55÷60°C. However the assumption of hot water temperature on the level of 45°C by calculations has a certain justification. Because on the one hand it is possible to achieve higher efficiency of heat pump or solar collectors, on the other hand it should be remembered, that the optimum temperature of mixed water for bathing and the wash is 38°C [109]. On the other hand, in water at ca. 37÷45°C the bacterium Legionella, which is harmful to health, develops fastest [110-112]. Taking this into consideration, the hot water temperature should be increased periodically above the value of 70°C. This makes it possible to use thermal disinfection in the installation to minimize the possibility of growth of the bacterium Legionella. The condition focusses attention on the necessary surplus of heat power by selection of heat sources.

After the definition of the required thermal power, which must be delivered by the heat source, one should define the kind of heat pump installation:

- monovalent,
- bivalent - alternative,
- bivalent - parallel,
- bivalent - partially parallel.

Most often in the case of air-water heat pumps the bivalent option is recommended, because in this way the device will not overestimate the heat demand appearing in the most part of the heating season. Therefore, the so-called bivalent point may by defined, which is expressed by a certain value of outdoor temperature (for example t_e = -8°C), at which the heat pump is supported by an additional heat source, among other things in the form of electric heating.

In the case of 'single source' exploitation, the heat pump works independently, so it is not supported by any additional heat sources, and must therefore cover the total project thermal load of the building. The monovalent installation assures low energy consumption which is used only to supply current pumps and compressors in the case of compressor devices. An advantage of this option is that the heat pump works in intervals, in which the bottom heat source is loaded to a lesser degree. One frequent recommendation is even to overestimate the heat pumps, which extends the pause of the device. Disadvantages of such solutions are the high investment costs of the heat pump and the installation of the bottom heat source. Another method of exploiting the heating installation with a heat pump is the bivalent system in which the heat pump works with a second heat source, which can be, for example, a coal, gas or oil boiler. In this system the heat sources can work parallel or alternatively. In case of the bivalent - alternative system, the main source is the heat pump which at the top heat demand, when its heat power is insufficient, is switched off. Then the heating of medium is taken over by a second heat source (Figure 5a). However, in the bivalent parallel system two heat sources work. The heat pump covers the total projected thermal load to a certain outdoor temperature, the so-called "temperature of the bivalent point of switching" on the second heat source. From this temperature the second heat source begins to work and the two heat sources work in parallel (Figure 5b).

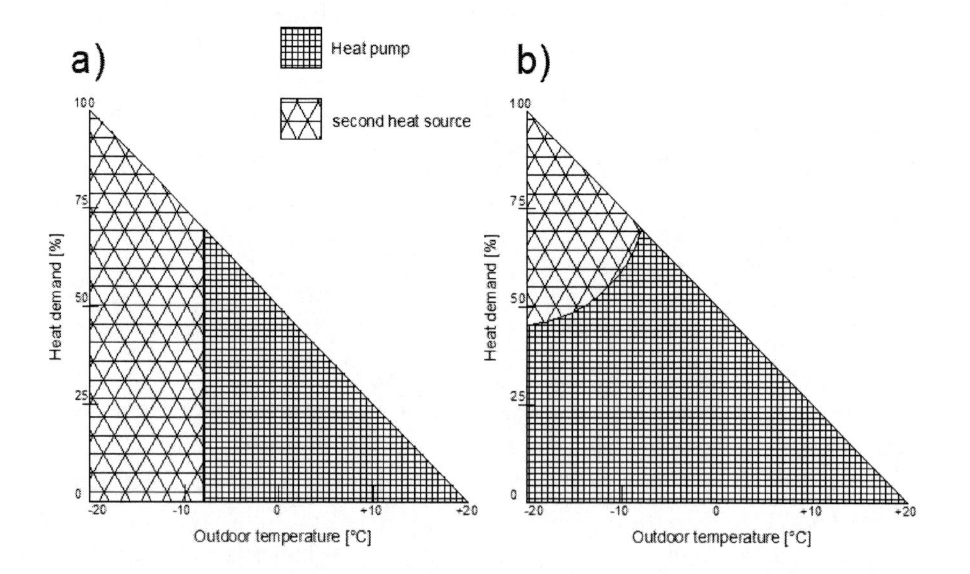

Figure 5. Types of heat pump operation: (a) bivalent - alternative; (b) bivalent- parallel.

In the case of a bivalent, partly parallel system, the heat pump covers the total projected thermal load to the certain temperature at which point the second heat source joins in and both work in parallel. However, on further lowering of the outdoor temperature the disconnection of the heat pumps follows and then only the second heat source works.

Typically, bivalent heat pumps are sized for 20÷60% of the maximum heat load and meet around 50÷95% of the annual heating demand [113].

Besides the above, one can include the so-called "monoenergy exploitation", in which, besides the heat pump, an additional heat source (for example, an electric heater) is applied. The additional heat source uses the same kind of energy as the compressor, in the case of compressor devices. The thermal balance of the building is calculated for the outdoor temperature, which depends on the climatic zone and in Poland amounts from -16 to -24°C. However, since the projected outdoor temperature appears only sporadically, one can purchase a device with lower power than that resulting from calculations, cut investment costs, and the top heat demand will be realized by use of an additional heat source. Such a solution allows users to decrease the capital costs, because the heat pump with lower heat power will be cheaper. In this situation the installation of the bottom heat source will have lower costs while maintaining the energy efficiency at a satisfactory level.

It should be remembered, that the greater the amount of heat it is possible to absorb from the bottom heat pump source, the lower will be the electrical energy required to drive the compressor, given the work parameters of the device. Therefore, it is very important to correctly choose, and then dimension the bottom heat source for the compressor heat pump.

In general, the bottom heat source of the heat pump should possess the following features:

- large heat capacity,
- a possibly high and constant temperature,
- lack of pollutants causing corrosion of installed components,
- easy accessibility and low cost of the execution and later exploitation of its installation.

Most often, renewable energy sources (external air, the ground, the solar radiation, the surface water, the ground water), as well as waste heat (air, gases, sewage, water cooling) are applied as the bottom heat source.

It is important, in selecting the bottom heat source for a heat pump, to take into account the compatibility of its temperature with the required power of the heat pump in working conditions. This has special meaning in the case of devices supplying heat to central heating installations, because within a period, when the demand is greatest on the thermal power, most of the bottom heat sources are characterized by the lowered temperature. In some cases, this temperature is lowest by the top demand on the thermal power for the heating. From this it results that most compatible bottom heat sources are: underground waters, the ground, surface waters, and lastly outdoor air.

Atmospheric air is one of most accessible renewable energy sources that can be universally applied. However, one of basic disadvantages of taking heat from the air is the seasonal variability of its temperature. A typical example can be seen for the weather conditions of Warsaw city (the capital of Poland) in Figure 6. In addition, air has a low heat capacity, making it necessary to press considerable air volumes through the evaporator. This causes an increase in the energy consumption (ventilators). Further, air is characterized by low value of the surface film conductance, which forces the use of larger heat exchange surfaces.

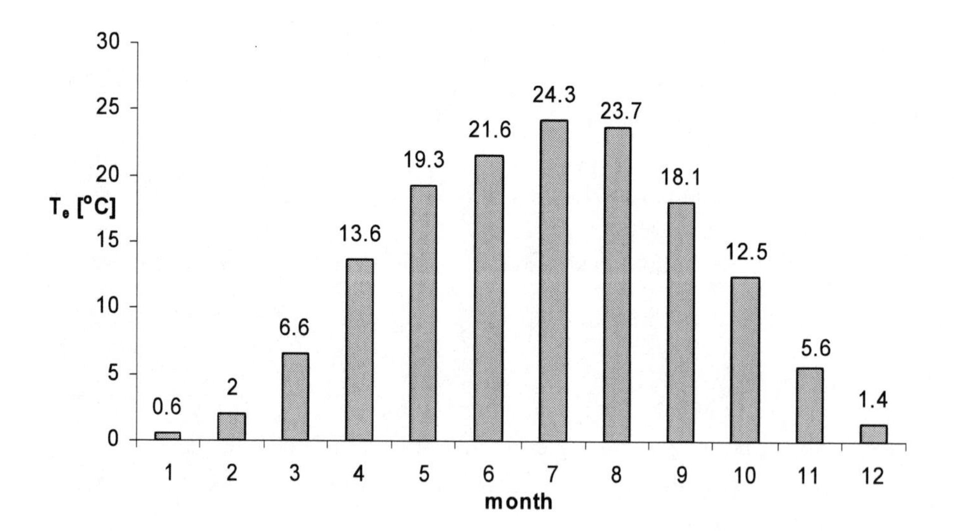

Figure 6. Average monthly values of outdoor temperature for Warsaw 1982-2010 [114].

Further, at outdoor air temperatures below 0°C the possibility of formation of a rime layer on the evaporator surface exists. This can cause an increase in the thermal resistance of the evaporator, and also make impossible the air flow through it. Defrosting is controlled usually automatically and is realized by way of decompression of a hot stream of the working factor to the evaporator instead of to the condenser, or else thanks to the electric air heater. For correct and economic working of the heat pump, the control set should be applied, which adapts the parameters of the heat pump to the external environment and parameters of thermal comfort inside the rooms.

Other solution are heat pumps of the type SPLIT in which the device to take heat is placed outside the building and is connected with the condenser of the heat pump by means of two canals in which the working factor is transported. This solution allows the removal of potential noise sources from the building.

Water is a low-temperature heat source, which is characterized by a considerably higher heat capacity than air and with a large coefficient of taking over the heat which allows the dimensions of heat exchangers to be reduced.

Nonetheless, in consideration of the small decrease of water temperature in the evaporator (about 4÷5°C), it is necessary to transport considerable mass streams of water, from which could be obtained on average 4500÷5900 (W·h)/m^3.

One can use the heat of surface waters, ground and deep-sea water. Wells tapping ground waters and deep-sea waters can be installed in Poland to a depth of 30 m without the agreement of the Main Institute of Mining [115]. Using ground and deep-sea waters for energy supply purposes is subject to the Water Law, which forces the preparation of the appropriate documentation and the installation of the low-temperature bottom heat source in compliance with guidelines concerning this type of device.

The temperature of surface waters is subject to the influence of changes in weather and climate. It can fluctuate from 3°C in winter-months to 23°C in summer-months, but the most often assumed values are in the range 5÷15°C. Using this kind of bottom heat source the possibility arises of obtaining the heat from the surface water at many sites along a river, or else from a reservoir, because the temperature in these sources is defined by direct heat exchange with the environment.

Currently the most produced water - water heat pumps, can work within a range of water temperatures in the bottom heat source between 7÷25°C, which limits the possibility of the use of surface waters within a period of the greatest power requirement (winter).

Anther solution is to obtain the heat from the ground water, whose temperature is more stable and fluctuates only within the range 7÷12°C.

The tap-well should have an efficiency not lower than 1.5÷2.5 m^3/h. The chilling of the heat medium could amount to 4 to 5ºC, and the chilled water is pumped up to the drain-well, from where it infiltrates back to the ground.

A next important problem is to define the distance between the tap well and drain-well, which can be carried out using Equation (33).

$$L = \frac{2}{\pi} \frac{\dot{V}}{i \cdot k \cdot H} \text{ [m]} \quad (33)$$

where:

$\dot{V}$ - stream of water volume, m^3/s,
k - speed of the ground filtration, m/s,
H - thickness of the water-bearing layer, m,
i - individual level drop of the water mirror (depression), ‰.

In addition, for dimensioning this kind of bottom heat source for a heat pump the following factors should be taken into account:

- results of the geohydrologic survey,
- temperature of ground water,
- stream of ground water supplied to evaporator,
- the level of ground water and the water-bearing layer,
- water chemical composition.

Additionally, the results of water analysis should be taken into account to minimize, as far as possible, such processes as: the growth of silt on the sides of the well, the precipitation of iron and its compounds, the corrosion of heat exchangers, because the large content of iron oxides and the magnesium in transported water can cause blocking of the drain-well. However, the high salinity of water increases the probability of damage to the evaporator as result of corrosion.

In the case when water does not ideally fulfil the properties given by the manufacturer, one ought to apply indirect circulation to the heat exchanger. Most often applied are dismountable heat exchangers produced from stainless-steel. The use of the indirect circulation stabilizes the heat exchange process, because heat is transferred more evenly than in the case of direct introduction of water to the heat pump. In this instance it becomes necessary to use an additional circulation pump. This will distort the flow in the indirect circuit, resulting in the fall of the heat pump efficiency.

The ground is characterized by a stable and quite high temperature. For this reason it is commonly used as the bottom heat source for heat pumps. Shallow geothermal ground source heat pumps have shown the most significant impact on the direct use of geothermal energy [116,117].

The authors [118] proved that the energy demand of the system, with a vertical ground heat exchanger (GHE) of parallel connection coupled to a heat pump system for air conditioning a public building, was significantly lower compared to that of conventional heating and cooling systems. The primary energy required by the system for heating was estimated to be lower by 45% and 97% as compared to that of air-to-water heat pump based and conventional oil boiler, respectively. In cooling mode the relevant differences were estimated at 28% and 55% for air-to-water and air-to-air heat pump based systems.

Heat accumulated in the ground is received by means of indirect systems in the form of phrenic heat exchangers, which can be made as horizontal, as well as vertical. Pipes between the heat pump device and the bottom heat source (the ground) might have contact with external air at low temperature, and for this reason they are usually used as the heat medium. Ecological liquids characterized by a low temperature of coagulation are favoured. They can be aqueous solutions with propylene or ethylene glycol.

The accumulation of heat in the ground and the influence of weather conditions on thermal parameters are evident to a depth of about 10 m. Below this depth, the value of the ground temperature is on the level of the annual external air temperature. At these depths in the range from 1 to 2 m the ground temperature is subject to certain oscillations, from about 5°C in January to about 17°C in July. These oscillations can be different for different kinds of grounds, because they are relative to humidities, the specific heat, as well as to the coefficient of the heat conduction of the ground (λ_g). Some typical values of the latter are shown in Table 10.

The ground horizontal exchanger is made usually from plastic pipes. Pipes most often are placed at a depth from 1.0 to 2.0 m. Generally they should be situated about 30 cm below the level of frost penetration in the ground. One should not lay sections longer than 100 m since the resulting large resistances to flow will cause greater consumption of electricity by the current pump. Individual circuits of the horizontal exchanger should have equal lengths, so that the flow resistances in them are identical. This will ensure equal flow in each section sections and will result in the uniform reception of heat from the total surface of the ground exchanger. Circulations are connected to distributors, which, for the purpose of making deaeration of the exchanger possible, are placed higher. The arrangement in the ground can be in different configurations: as row, parallel, pipe coil (Figure 7a) and spiral systems (Figure 7b) [119].

Table 10. Values of the heat conduction coefficient of the ground (λ_g).

Type of ground	λ_g [W/(m·K)]
Sea saturated sand	2.44
clay ground	2.33
moist ground	2.10
water saturated sand	1.88
Sea-sand with humidity of 20%	1.76
meanly moist ground	1.40
sand with humidity of 20%	1.33
sandy ground	1.16

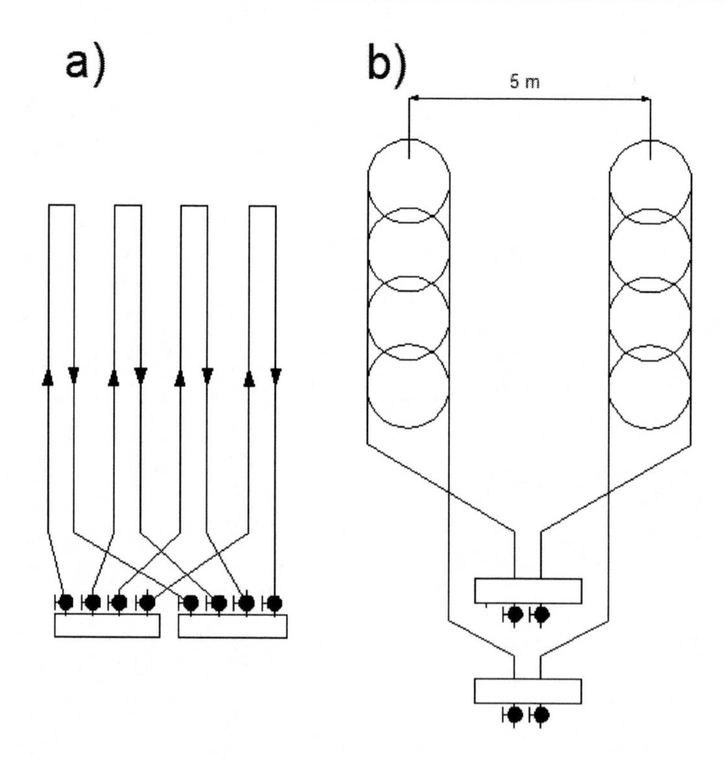

Figure 7. Scheme of the horizontal ground heat exchanger: (a) pipe coil; (b) spiral.

During the designing stage of the ground horizontal heat exchanger data about the geological structure of the ground and the average temperature distribution in the ground are indispensable, to define the required length of pipes for heat exchanger (L). The required length of the horizontal exchanger can be defined by use of Equation (34):

$$L = \frac{Q_o \ln \frac{4 \cdot h}{D_{ex}}}{2 \cdot \pi \cdot \lambda_g \cdot (t_h - t_{supply})} \text{ [m]} \quad (34)$$

where:

Q_o - thermal power received from the ground, W,

h - depth of arrangement of ground exchanger, measured from the ground level, m,

λ_g- heat conduction coefficient of the ground, W/(m·K),

D_{ex}- outside diameter of ground exchanger pipe, m,

t_h- ground temperature at depth (h) with allowance for ground cooling as a result of the heat collection, °C,

t_{supply}- temperature of the heating factor supplying evaporator of heat pump, °C.

After correct dimensioning of the bottom heat source, an important problem is ensuring the correct execution of the horizontal exchanger, because this influences its later exploitation. Pipes should be arranged at a distance of 0.5÷1.0 m from each other. However, if this is not possible, then the difference of depth between neighbouring pipes should be kept in the range of 0.4÷0.5 m. It is recommended to make each section from a single pipe, in order to eliminate connections, which are one of most frequent places of leakage and later failures. By realisation of the horizontal heat exchanger, the ground surface should be provided to build it. This required surface is strictly connected with the required thermal power, which has to be taken from the bottom heat source. On the ground over the heat exchanger there should be no plants which have an expanded and deep radicular system, because they would be able to damage the pipe system, which takes heat from the ground. Additionally, in consideration of the thermal regeneration of the heat exchanger, thanks to heat gains from the insolation and falls, these ground surfaces should not be cover with concrete.

In the case when investor does not have the required surface to construct a horizontal ground heat exchanger, an alternative is the vertical ground heat exchanger which can be made in the forms:

- U-type,
- with backward flow,
- with concentric flow,
- spiral.

Nowadays the most often constructed exchangers in U-shaped form are made from pipes with diameter from ¾" to 2". In consideration of the costs of the bore hole more and more often ground already found in the form of double U-type are prefered.

To define the amount of heat, which is possible to receive per m of the bore-hole, the values summarised for different types of the ground in Table 11, can be used. The values apply to the following conditions:

- depth of sounds is in the range from 40 to 100 m,
- a minimum distance between sounds is kept:
  - 5 m, by depth of sounds is in the range from 40 to 50 m,
  - 6 m, by depth of sounds is in the range from 50 to 100 m,
- The working period of heat pump with the maximum thermal power does not exceed 1800 h in the year,
- sound is made in the system with the double U-type sound with pipes Ø 25 or Ø 32 mm.

Table 11. Approximate values of obtainable heat per m of the bore-hole constructed from the double U-type sound [120]

Characteristic of the ground	Received thermal power (q_e), W/m
Ground with disadvantageous thermal properties $\lambda_g < 1.5$ W/(m·K)	20
usual bed-rock and water saturated deposits $\lambda_g = 1.5 \div 3.0$ W/(m·K)	50
solid rock with large thermal conductance $\lambda_g > 3.0$ W/(m·K)	70
large ground flow in sands and gravels	80÷100

Similar values were observed by another research group [94], who stated that, at an average depth of 150 m the thermal yield is about 50÷70 W/m. The depth depends on the heat load, the thermal conductivity of the ground, the natural temperature in the ground, the ground water level, and the distance to other ground-source heat pump systems.

If, in the vertical profile of the ground sound, subterranean waters do not appear, the heat exchange between sound and the ground is an unestablished process. In this case the thermal resistance of the ground is a time function. On the other hand, if the ground probe on its length has a contact with subterranean waters, than the heat exchange between the probe and the ground may by considered as a conditioned established process. That's why, during drillings, the length of probes can be verified, on the basis of the information concerning the geological basis given by the geologists who made the bore hole.

Another method to improve the heat pump performance is the integration of a heat pump with solar technology in the form of a hybrid system, as presented in the studies [121-123], and in [124] especially for heating-dominated buildings. Also, decreased use of electricity could be achieved by using a system in which the solar collectors produce domestic hot water during summertime and recharge the borehole during winter [107].

The upper source, which is recommended for the heat pump is low-temperature plane heating, especially floor heating. Heat transfer by radiation, which is the main mechanism of heat exchange in panel heating systems, to people, surrounding objects and partitions is more economic than heating up the total air volume by way of convection, which fills the heated room. This solution allows one to lower the average temperature of the air in the room by ca. 1÷2°C, resulting in a diminution of the energy consumption of about 10%. The decrease of the required temperature on supply of heating system contributes to an increase of the heat pump efficiency (*COP* coefficient), the diminution of heat losses on transfer and to improvement of thermal comfort conditions [125,126].

Besides the experimental research, the IEA Research Project [127] showed that persons living in buildings equipped with low-temperature heating systems were more satisfied in indoor environment than persons living in buildings with high-temperature heating systems. Hasan et al. [128] shows that systems with floor heating consumed about 50% less primary energy than systems with convective radiators in rooms.

Besides, the floor heating system is characterized by an almost ideal air temperature distribution in the vertical profile. Most people agree that the most comfortable vertical profile of temperature within a room is when the air at floor level is warm and with increasing height becomes cool(er). Fundamentally, panel floor radiators consist of four basic layers:

- anti-moisture isolation,
- thermal isolation,
- heating panel,
- cement floor.

The thickness of these layers is related to the construction of the floor or the kind of the ceiling.

The room to be heated determines the choice of the suitable arrangement of pipes in the panel floor radiator. In practice several systems are accessible. Most often applied is the loop system (Figure 8a), whose main advantage is the uniform distribution of the floor temperature. When the heated room is heterogeneous and one of its partitions clearly has greater heat losses, then the meander system can be applied (Figure 8b).

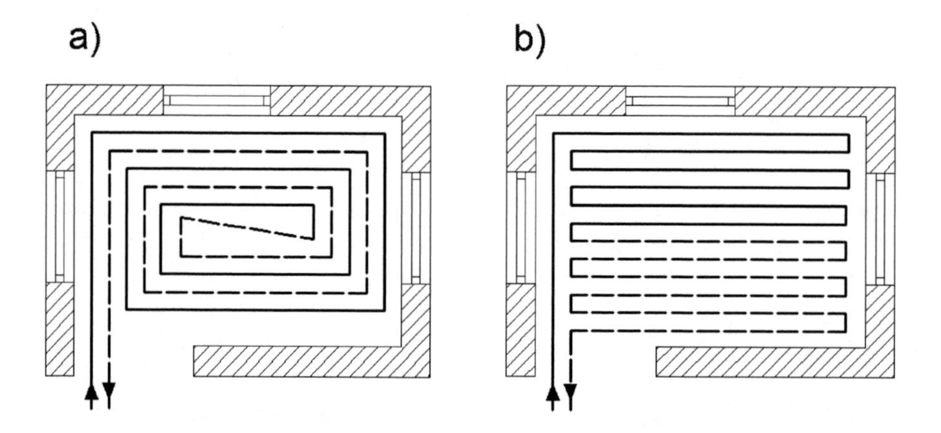

Figure 8. Types of piping schemes used in panel floor heating systems: (a) loop system; (b) meander system.

4.4. Application of Solar Thermal Systems

Another possibility to decrease the energy consumption is to use conventional systems which cooperate with installations using solar radiation energy. This concept has been implemented for some time in Greece [129] and other countries [130].

Solar energy is nowadays considered as one of the most environmentally friendly technologies, among other things by virtue of its many obvious advantages [131-133]:

- reclamation of degraded land;
- no emissions of greenhouse (mainly CO_2, NO_x) or toxic gasses (SO_2) during operation;
- reduction of transmission lines from electricity grids;
- improvement of quality of water resources;
- increase of regional/national energy independence;
- diversification and security of energy supply;
- acceleration of rural electrification in developing countries.

The sun emits electromagnetic radiation in all directions. At the outer stratosphere normal to the flow of solar radiation the flux density amounts to G_{sc}=1367 W/m^2; this is known as the solar constant. On the Earth's surface, owing to absorption and dispersion processes and by the influence of cloudiness, the solar radiation level is lower than the solar constant. Therefore, the total radiation arriving at the Earth's surface is a result of direct radiation, dispersed radiation and reflected radiation.

The direct radiation (I_b) is short-wave radiation with the direction of the rays dispersal in a straight line from the Sun to the active surface of a solar collector.

The dispersed radiation (diffusive) – (I_d) - arises as a result of the reflection of solar radiation from dust- and gaseous pollution in the atmosphere. This sun dispersion is the reason for the blue colour of the sky. Energy gains are in reality smaller than gains resulting from direct radiation [134].

For example the solar radiation density in Poland amounts to:

- 950÷1100 kWh/(m^2·year) [135];
- 950÷1150 kWh/(m^2·year) [136];
- 950÷1250 kWh/(m^2·year) [137].

Annual values of solar radiation density after factoring-in of energy losses, resulting from absorption and dispersion in the atmosphere, amount to 780÷1050 kW·h/m^2, depending on the region of the country. However, it should be remembered that up to 80% of the total annual sum of solar radiation arriving in Poland is in the period from April to September, and only 20% in the heating season [138].

Meteorological conditions, even those presented in Table 12 (including cloudiness), influence the possibility of solar energy use. Whenever clouds appear, direct radiation weakens, hence the maximum of direct radiation can be expected only during completely cloudless days.

Table 12. Dependence of total solar radiation intensity on weather conditions [134]

Weather conditions	Total solar radiation intensity [W/m^2]
Cloudless, blue sky	1000
Sun shows through clouds	600
Sun as white pane	300
Cloudy, winter day	100

During the summer season, when stormy weather with strong vertical development of clouds occurs, the greatest disturbances appear in cool and transparent air masses. In such situations, the strongly variable direct radiation overlaps the strongly variable dispersed radiation. For this reason, the temporary, total value of both radiation components can considerably exceed values registered on cloudless days.

By the very favourable arrangement of high storm clouds, the radiation rate on the ground level may even reach values above 1300 W/m^2 [139].

The study [140] describes many ways of using solar energy.

Solar electrical direct conversion systems (efficiency ca. 17%) are not so effective in comparison to solar thermal conversion, (ca. 70%) [141]. Taking this into account, hot water preparation for residential buildings has attained the widest popularity globally. This is connected with the asthetics of their construction, which does not demand the usage of advanced technologies and expensive materials. In fact, the cost of hot water preparation with solar collectors is on average about 1.84 times smaller than using only a gas boiler. The greatest difference in costs is in the summer months when the cost is ca. 2.98 times less in the case of the solar collector installation [142].

Development of the solar water heaters has mainly concentrated on their appearance and the types of material used in their construction.

Basically, every solar hot water installation consists of:

- a solar collector,
- a storage tank,
- a transporting system of heating medium in section between solar collector and storage tank,
- a control and steering system.

Additionally, control elements such as safety valves, an expansion tank and a deaerator are frequently used.

There are many variants of hot water installation, which differ in their method of transporting the heating medium from the collector to the tank and the interconnection between individual elements. Some of these variants are presented in Figure 9.

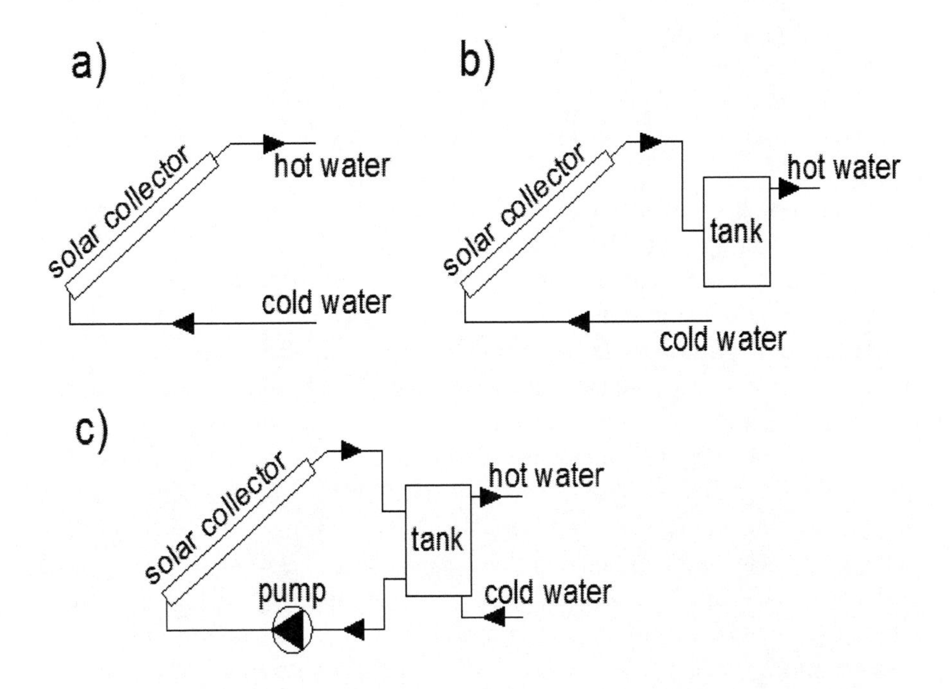

Figure 9. Some possibilities of solar hot water preparation: (a) direct without tank; (b) direct with tank; (c) direct with water flow forced by circulation pump.

Based on the circulating flow type and the presence or lack of a solar collector-storage tank, these installations can be divided as follows [143]:

- active installations, with forced flow, where the flow of heating medium is caused by circulation pump;
- thermosyphon installations, passive (without a pump), with the natural induced flow by differences between the densities of heating medium in the solar collector and storage tank. For example in the study [144] the authors evaluated the thermal performance and heat removal efficiency of twelve different types of thermosyphon solar water heating systems and suggested a criterion of modified efficiency of η=0.41. Also, Abreu and Colle [145] experimentally analyzed the thermal behaviour of two-phase closed thermosyphons with an unusual geometry characterized by a semicircular condenser and a straight evaporator;
- installations without the flow of heating medium, termed storage collectors; in these systems the reservoir is at the same time a solar energy collector.

Most often applied in Poland (temperate climate zone) are solar hot water preparation systems with the forced flow and with an exuded primary circuit (Figure 10).

In the primary circuit circulates a non-freezing mixture, necessary because negative ambient temperatures may occur.

In the case of an indirect circulation system of energy collection, the heat is returned to the heated up water by use of a heat exchanger. The heat exchanger is situated in most cases inside the storage tank, but it may also be an external device. The system with the external heat exchanger is applied by use of the same solar collectors by several heat receivers or in hybrid installations.

The control of heating medium flow in the collector circuit is realized by a special control system. These are most often threshold control units with two temperature sensors. One of temperature sensors is located at the bottom of the storage tank near to the outlet from the heat exchanger. The second temperature sensor is located near the connection ferrule of the heating medium leaving the solar collector. When working, the temperature difference between these sensors is approximately equal to the increase of the temperature of the medium in the solar collector. A pump is usually switched on after the a certain temperature difference (ΔT_{ON}) between these two sensors is exceeded. It is switched off, when the temperature increase of water in the solar collector falls below the second liminal value (ΔT_{OFF}).

The value of temperature difference at which the pump is switched on, should not be too small, because this will cause frequent switching on and off of the system. It is not recommended to set smaller temperature differences for switching on the pump than ΔT_{ON}= 5°C.

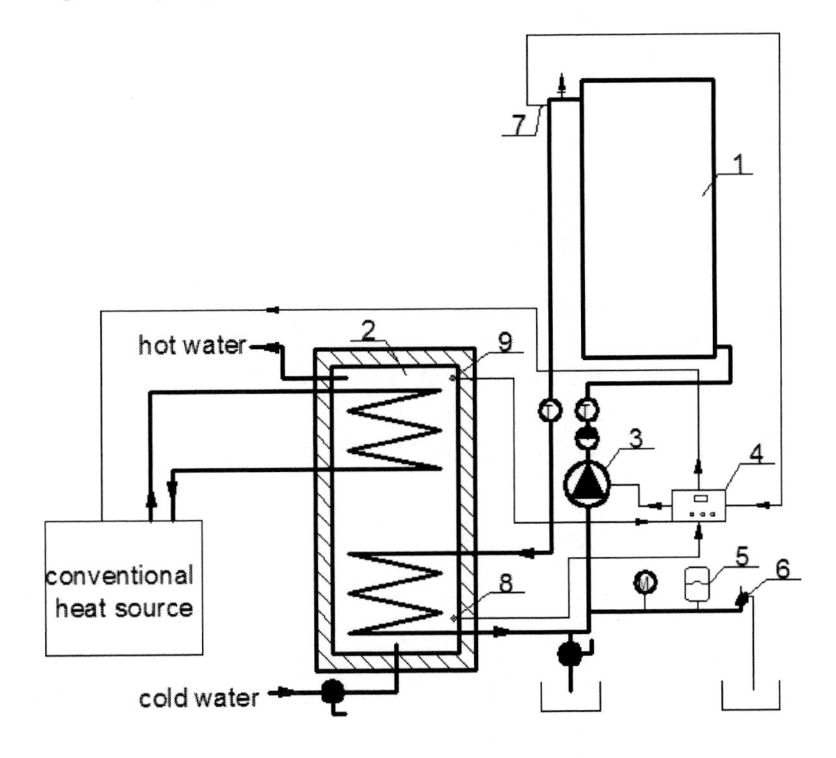

Figure 10. Scheme of a solar hot water preparation system using forced flow in indirect circuit to: 1- solar collector; 2- storage tank; 3- circulation pump; 4- control box; 5- expansion tank; 6- safety valve; 7- temperature sensor of the heating medium on the collector exit; 8- temperature sensor of water in bottom parts of the storage tank; 9- temperature sensor of water in upper parts of the storage tank.

In this case, the controller of the solar circuit should fulfil the following functions:

- switch on the circulation pump in the solar collector loop, when the difference between the temperature of the heating medium near the outlet from the absorber near and the temperature in the lowest part of the storage tank exceeds a given liminal value, (ΔT_{ON}),
- switch off the circulation pump, when the above temperature difference falls below the second, lower value of the liminal temperature (ΔT_{OFF}),

- switch on the second conventional source, which will heat up water in the highest parts of the storage tank, if its temperature cannot be assured by the solar system and falls under the given value,
- switch off the circulation pump, when the temperature in the upper parts of the storage tank exceeds the maximum value (about 90°C), to avoid damage to the tank.

Another possibility of controlling the work of solar thermal systems with forced flow is to change the rotations (rpm) of the circulation pump (the proportional control), in other words to change the mass stream circulating in the solar system. However, it should be remembered that the total energy gained per day from the collector by proportional control is smaller than by liminal control. This is caused by the diminution of the coefficient of heat transfer from the collector together with some loss of the mass stream, flowing through the collector [146,147]. However, an advantage of this control is the possibility of obtaining constant increases of the heating medium temperature in the collector independently from that of the value of the solar radiation reaching the collector, which allows the supply of hot water in the morning hours. Most often this type of control is used in „low flow" installations, in which only low mass streams of the factor flow.

Under the concept of solar systems efficiency several dimensions, are commonly used, namely [148]:

- calorific effect expressed in the energy unit per m^2 of the solar collector surface (MJ/m^2, $kW \cdot h/m^2$); daily, monthly, seasonal or annual efficiencies are frequently in use,
- coverage level of energy demand (*SPT*) referred to hot water preparation, expressed in percentages, considered for the summer-, winter- and yearly periods; can be estimated by use of Equation (35);
- efficiency of solar thermal energy conversion, expressed in percentages and refered to twenty-four hours, the month or the year.

$$SPT = \frac{E_{solar}}{E_{solar} + E_{conventional}} \cdot 100\% \tag{35}$$

where:

SPT - coverage level of energy demand for hot water preparation by the solar installation, %;

E_{solar} - heat obtained from solar collectors for hot water preparation, kWh/a;

$E_{conventional}$ - additional conventional energy necessary for hot water preparation up to the required temperature, kWh/a.

If the objective performance of a given solution is required, then the simultaneous calorific effects or the conversion efficiency and the level of coverage of thermal needs must be provided. These assumptions avoid the appearance of differences when comparing the installation, because that with the larger area of solar collectors in reference to thermal needs will return the higher degree of cover, but the individual calorific effect will be then small. Similarly, when the area of the collectors is too small, then the individual calorific effect will be high, but the level of coverage of heat demand will be low.

If we are concerned about the solar system efficiency, then it could be different depending on the location of the measuring node, which can be installed [148]:

- at the level of the solar collectors, then the quantity of heat taken from the absorbing units may be obtained;
- at the level of the storage tank, this means the measurement of the quantity of the energy delivered to the storage tank; this takes into account heat losses in the section solar collector - storage tank;
- at the level of hot water consumption, when it is required to separate from the energy raised together with hot water, the amount of energy which was supplied by the solar installation.

Additionally, evaluation of solar thermal system depends on the methods of received solar energy, because according to the classical concept [149], the energy delivered to the storage tank when the temperature exceeds the nominal level, is not included in calculation. A basic element in all solar systems is the solar collector. Solar collectors depending on the kind of heated up factor; they can be divided into:

- collectors to heat up water,
- collectors to heat up air.

Solar collectors used to heat up water - liquid collectors - may be classified as follows: flat plate collectors and vacuum pipes collectors.

One type of solar collector consists of the storage system, which integrates the collector and storage tank into one piece of equipment [150]. Thus, it differs from more conventional types by its simplicity of structure: no divided collector, no storage tank, no connection pipes and only a small area is needed for installation. The instantaneous efficiency of the storage collector was in the range 0.4÷0.78 compared to 0.21÷0.35 for the conventional one.

Generally, solar energy collectors are constructed of the following:

- absorber, which should be include the following features:
  - short time of heating up,
  - constant temperature,
  - high resistance to corrosion,
  - effective heat transfer to the liquid heating medium,
  - low flow resistance.
- solar pane of glass, which should allow solar beams to pass easily and simultaneously maintain the radiation of the collector's absorber and heat losses by convection at the lowest possible level. This pane is initially thermally tightened and is characterized by a strongly reduced amount of iron, therefore its degree of transmission is around 92% [134];
- an enclosure, which consists of the collector frame and the heat insulation; this isolation should maintaing minumal heat losses in relation to each side of the collector.

The efficiency of solar collectors, which is the ratio of the usefully effective energy to the solar radiation energy, can be enumerated by use of following dependences (Eq. 36) [151].

$$\eta = \frac{Q_u}{I \cdot F_{collector}} = \frac{\dot{m} \cdot c \cdot \Delta T}{I \cdot F_{collector}} = (\tau \cdot \alpha) - U_L \cdot \left(\frac{T_A - T_O}{I} \right) \tag{36}$$

where:

η- temporary solar collector efficiency,
I- the intensity of total solar radiation falling on solar collector surface, W/m^2,
$F_{collector}$- solar collector surface, m^2,
Q_u- effective power, W,
$\dot{m}$ - mass-flow rate of liquid by solar collector, kg/s,
c- specific heat of heating medium, J/(kg·K),

ΔT- increase of temperature of heating medium, K,
U_L- coefficient of heat losses from absorber; usually $U_L = 3\div10$ W/(m^2·K),
τ- transmission coefficient of absorber cover; $\tau = 0.8\div0.9$,
α- coefficient of absorptivity of absorber cover; $\alpha = 0.6\div0.9$,
T_A- absorber temperature, K,
T_o- ambient temperature, K.

Higher efficiency of solar collector may be obtained by way of:

- greater flow rate of heating medium through the solar collector, because the temperature difference between the absorber and heating medium increases, which simplifies the process of heat exchange in the absorber; additionally, growth of flow rate favours increased turbulence which also intensifies the heat exchange,
- use of an additional flat reflector by the solar collector,
- use of double glazing or a selective cover,
- increase of thermal insulation thickness.

The effect of the annular space between the riser tube and absorber plate also influences the efficiency of the solar collector, what has been analyzed experimentally and theoretically by many researchers [152,153]. The effect of dust accumulation on the glass cover of solar collector has been investigated experimentally with different tilt angles by Hegazy [154] and by Soulayman [155]. The fractional reduction in glass transmittance depends on dust deposition in conjunction with plate tilt angle, exposure period and site climatic conditions. The 0° tilt angle is most contaminated with a mixture of coarse and fine dust particle and 90° tilt angle has least amount of dust accumulation.

An important problem in order to achieve the maximum efficiency of the solar radiation for hot water preparation is selecting an optimal location of the solar collector. Optimal angles of the stationary solar collector surface in relation to the ground level for Polish weather condition are summaried in Table 13.

Another very important aspect, is the suitable location of solar collector in relation to compass orientation. As far as the southern orientation of the roof for the solar collectors installation is universally accepted, little attention has been paid to roofs with other orientations, and how these can be usefully used. This is comprehensively discussed in the study [157]. It was ascertained, that for flat and vacuum solar collectors, the level of coverage of energy demand by solar installation for hot water preparation decreased maximally by about 4% in relation to the maximum value, in the range of β angles =10÷60° and deviations from south into the eastern and western side of about 45°.

Table 13. Values of optimal angles of stationary solar collector in relation to the ground level (β) [156].

Period of time	Optimal angle (β) [°]
January	75
February	65
March	55
April	40
May	25
June	25
July	20
August	35
September	50
October	65
November	70
December	80
Summer season	30
Winter season	70
Total year	40÷45

Of course, there are so-called tracking solar collectors, which can be divided into:

- solar collectors with a vertical rotation axis, these are horizontally tracking collectors, with a carousel type of rack,
- solar collectors with horizontal rotation axis - vertically tracking solar collectors,
- solar collectors with oblique rotation axis,
- solar collectors with tracking in two axes.

These types of solar collectors track the Sun position and, depending on the degrees of discretion aim to diminish the angle between the normal straight line to the collector surface and the actual direction of sun's rays. Thanks to these functions an efficiency increase of solar energy use of about 20% in relation to stationary solar collectors can be obtained [156]. Comparing average annual values of (*SPT*), for solar collectors of different mountings (Figure 11) located in Poland, it can be seen that the greatest values are obtained from solar collectors tracking in two axes. However, bearing in mind that these differences are not so large (all ca. 60%), before the ultimate decision concerning the choice of mounting it is wise to carry out a simple economic analysis of the profitability of the available options.

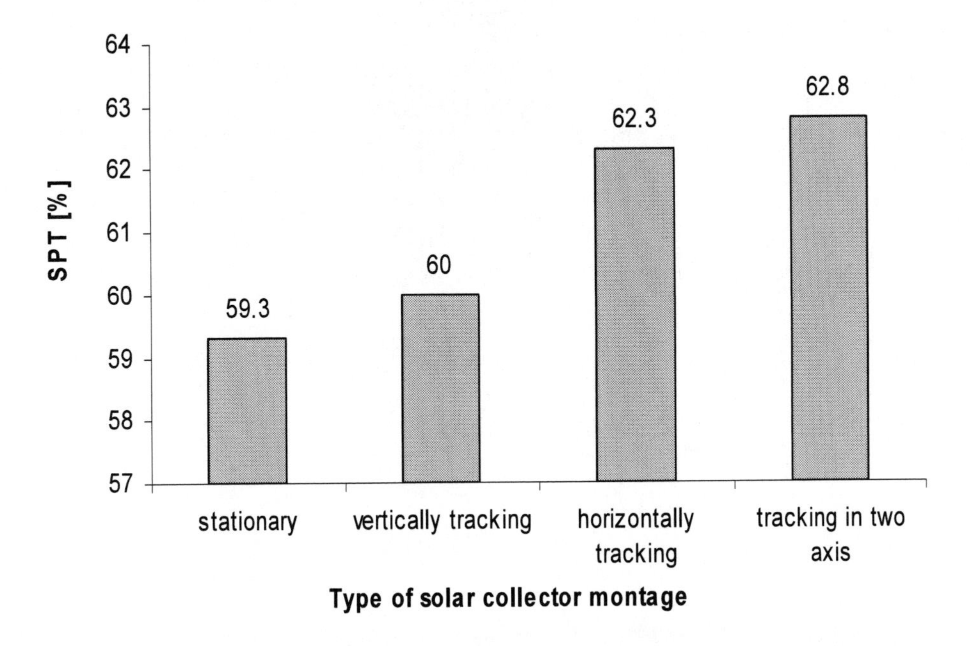

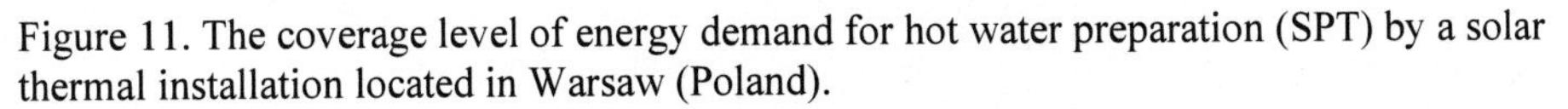

Figure 11. The coverage level of energy demand for hot water preparation (SPT) by a solar thermal installation located in Warsaw (Poland).

The indicator (*SPT*) depends also on the temperature of hot water, selected by the user of the installation (Figure 12). Here it is proper to set the lower temperature of hot water, because this may contribute to an increase of the indicator (*SPT*). However, by decreasing the hot water temperature, the enhanced growth of the bacterium Legionella should not be overlooked.

Besides, the *SPT* value also depends on the quantity of hot water consumption, (see Figure 13).

Another approach for the analysis of the performance of solar water heating systems, which determines the number of days in each month when solar heated water above a set temperature is available from the system, was proposed in the study [158]. Considering that the solar system efficiency is strongly dependent on weather conditions and the fact that, in the winter season, the cover of energy demand for hot water by solar collectors is not large, it is necessary to use second

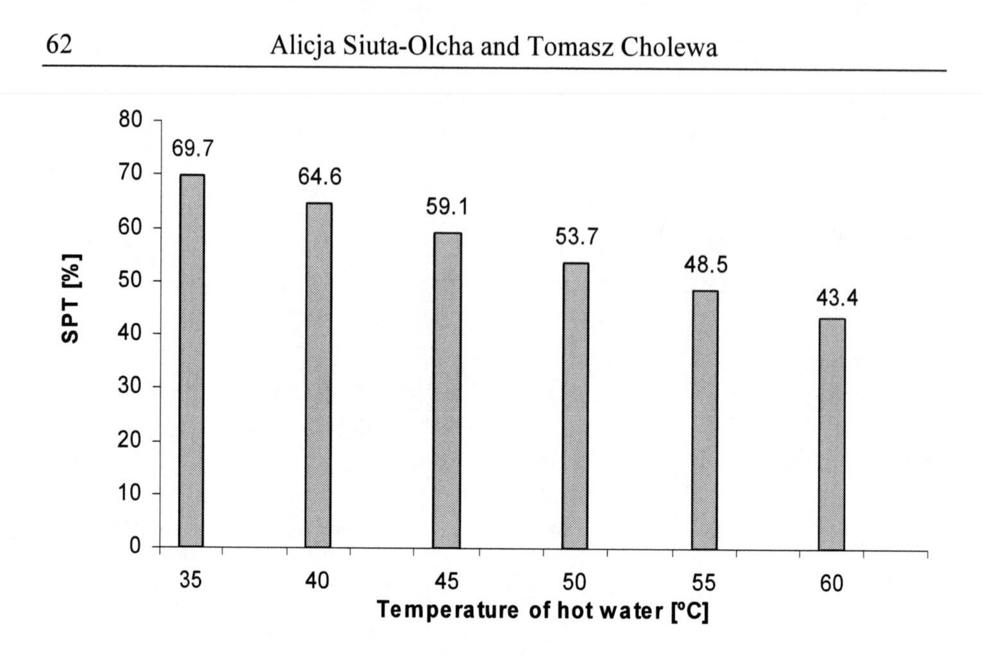

Figure 12. The *SPT* indicator in relation to temperature of hot water for single-family home in Poland.

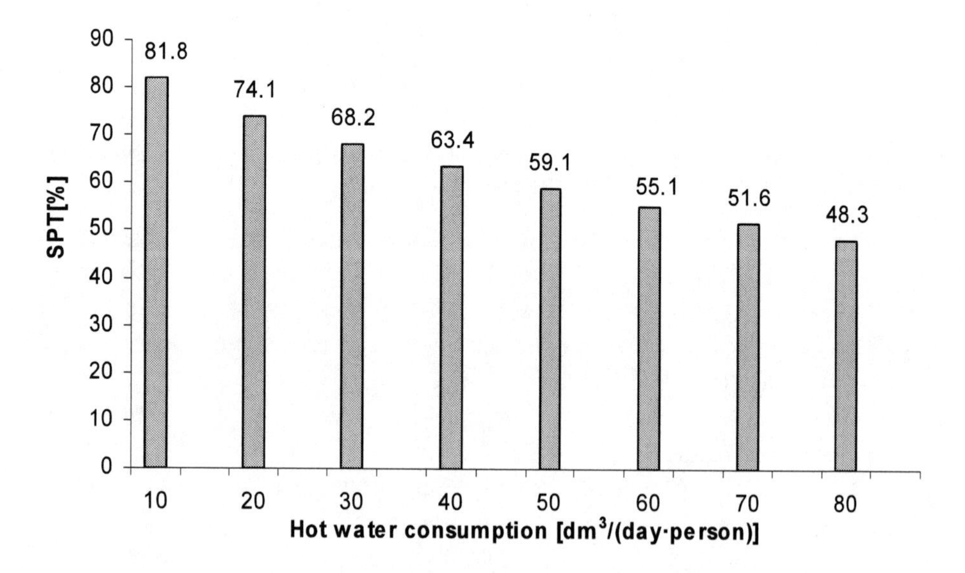

Figure 13. The *SPT* indicator in relation with quantity of daily hot water consumption per person, for a 4 person family living in a one-family home in Poland.

heat source. This heat source can be a central heating boiler (gas or oil), an electric heater, heat from a local district heating network or a heat pump.

Another, very often used indicator for comparing solar systems for hot water preparation, is the economic efficiency *SPBT* (Simple Pay Back Time), determined by use of Equation (37). This indicator is calculated from the start of investment until the moment at which the sum of savings, obtained as result of the realization of the investment, balances the investment cost. This indicator is very important, especially for investors, because of their wish that the installation costs amortize as soon as possible.

$$SPBT = \frac{N}{S_a} \text{ [years]} \tag{37}$$

where:

N- the investment costs in euro;

S_a- the sum of annual savings with relation to the reference case in euro/per year.

The investment costs by calculation of *SPBT* is the difference between the total investment on the installation with renewable energy source, and the investment on the reference installation, which is usually an installation with conventional fuel energy source. Values of *SPBT* received as a results of a simulation analysis of solar installation for hot water preparation in a one-family house depending on the kind of the conventional heat source are shown in Figure 14.

The *SPBT* indicator is usually calculated using computer programs, for example RETScreen, which was used very effectively to analyse solar water heating in Oman [159].

As previously mentioned, using solar installation provides essential environmental advantages (minimises combustions of fossil fuels and emissions of dusts and greenhouse-gases to the atmosphere). In order to define the quantity of substances emitted (or not emitted) to the atmosphere during the combustion of fossil fuels, one should, in the first instance, calculate the necessary amount of fuel per year or analysed period for hot water preparation, by use of Equation (38).

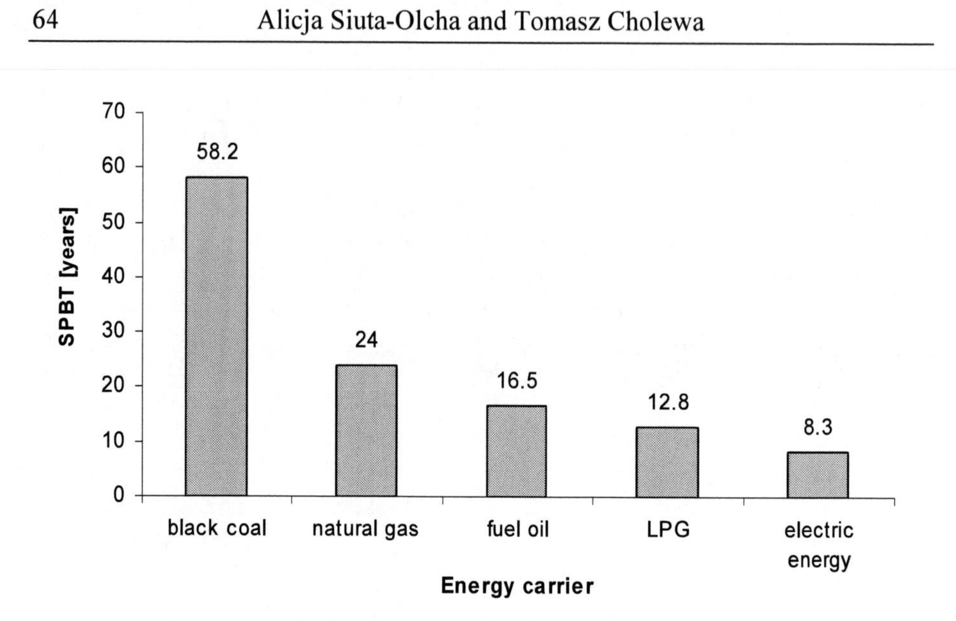

Figure 14. Simple Pay Back Time (*SPBT*) of invested capital for solar installation in one-family house depending on the conventional energy carrier.

$$B_a^{HW} = \frac{(24 \cdot 3600 \cdot Q_{HW} \cdot 36)}{Q_i^r \cdot \eta_{HS} \cdot \eta_{HW}} \tag{38}$$

where:

Q_{HW} - computational heat demand on hot water preparation, kW;
Q_i^r - calorific value of the fuel, kJ·(m^{-3});
η_{HS} - efficiency of heat source;
η_{HW}- efficiency of hot water installation running.

By use of the formula (Eq. 39) the average yearly emissions of individual pollutions from the heat source, can be estimated.

$$E_{emission} = B_a^{HW} \cdot u_{emission}^j \text{ [kg/year]} \tag{39}$$

where:

B_a^{HW} - yearly consumption of fuel;

$u^j_{emission}$- index of pollutant emission from the emission sources for NO_2 (u_{NO2}^j), CO (u_{CO}^j), CO_2 (u_{CO2}^j) and for dust (u_D^j).

In the study [160], the authors analyzed the realistic behaviour and efficiency of heating systems based on long term monitoring projects and hence showed that the potential of fuel reduction can be a maximum of double the solar gain due to a strong increase of the system efficiency.

Additionally, formation of wastes arising from rinsing and cleaning of fossil fuels is avoided. The degree of environment deterioration is also diminished, because fossil fuel production entails destruction of large areas of the earth's surface and changes in the structure of deep-sea waters (drilling, BP incident).

Another point worth mentioning in the estimation of (*SPBT*) is that additional costs arising from, for example, environmental remediation or the costs of pollution-related illnesses to society as a whole, are generally left unconsidered.

4.5. Control and Operation Systems in a Domestic Heating Installation

Operation and control systems could assist in mitigating energy use in residences by allocating the delivery of services by time and location more efficiently.

Considerable energy is wasted in delivering services such as heating or cooling unoccupied spaces, overheating/undercooling for whole-house comfort, leakage current, and inefficient appliances inefficiently to residents. The heater or air conditioner could be also switched off or simply turned down when residents are absent. Much of this waste is due to the fact that the level of energy control in homes is often out of date. For example only 28% of US homes had even a simple programmable thermostat in 2005, with only 58% (16% of US homes) actually utilizing the programmable aspects of their hardware [161]. Most homes rely on simple bimetallic strips or non-programmable digital thermostats. It is also very difficult for homeowners to decide on how to implement and evaluate energy saving strategies, because they receive very little information on energy use beyond their monthly or half-year meter readings.

One solution for efficient heating system control may be IT-enabled monitoring and control technologies, which are characterized by their ability to distribute and manage data and thereby allow more precise control of technological systems. These systems are able to refine the delivery of needed energy in both time and space – provide services exactly where they are required, when they are required, and in the amount that they are desired or needed. They are also capable of providing rapid feedback on energy usage to residents and could potentially break down energy consumption by appliance, room, time of day, and so on. This is in stark contrast to the traditional mode of feedback: a once-monthly bill featuring a single aggregate number for energy use.

For example one study showed that about 23% of residential primary energy is wasted due to inefficient energy services [162]. This was caused also by thermostat oversetting: variations by room or home area. There is a difference between the upstairs and downstairs temperature for a two storey home which is on the level of 2÷4.5°C [163] and only one central thermostat is localized in the building providing a signal to the heating control system.

The location of the room thermostat is very critical for the correct operation of the heating system. The thermostat should be located in a zone which is representative for the building. It should not be sited in southwards oriented rooms. Furthermore, since the wall on which the sensor is attached also has a strong influence on control system performance, internal walls are preferred [164].

A modified form of the interval heating option involves control of the heating system on the basis of the electrical energy consumption in the flat [165]. The temporary programs are arranged automatically on the basis of electrical energy and hot water consumption. In this solution the temperature in the room is maintained at a level dependent on the physical activity level of users and on the outside temperature. In comparison to continuous heating, it led to a decrease in energy consumption of about 19%, as well as being simple in operation. The latter is, in fact, its greatest advantage, as in the study [166] it was found that many persons, owning a heating system with standard control, lacked the ability to program it and neglected to use interval heating and temperature reduction overnight.

Another method of heating system control was presented in study [167], in which the author proposed a strategy which controls only indoor temperature and changes the set point according to additional measurements of indoor humidity to maintain thermal comfort. Non-linear compensation of outdoor temperature and wind speed was also introduced. The authors of [168] presented a model predictive controller, which used both a weather forecast and thermal model of a building to inside temperature control. This model can utilize the thermal capacity of a building and minimize energy consumption, even from the range of 17÷24% in comparison to a standard controller.

Cho and Zaheer-Uddin [169] state that the use of the control method of floor heating using the predictable hourly outdoor temperature, can give energy savings in the range of 10÷12% in comparison with standard control methods.

Generally, reduction of the internal temperature under computational control within an established time interval can be accomplished using one of the three options:

- complete shutdown of the heating installation (shutdown mode),
- decrease of the installation power depending on the outside temperature,
- regulation of the thermal power (internal temperature reduction mode).

Internal temperature reduction in residential buildings is generally carried out at night, so as not to decrease thermal comfort. However, in one-family buildings the temperature reduction can be applied additionally in daily hours, when the inhabitants are out of doors. This kind of saving can be obtained also in public buildings (offices, schools) and industrial buildings, because their users generally remain within the building only for part of the day and most often not at weekends. In research carried out in Bialystok (Poland) [170], where the internal temperature in a building was reduced from 20 to 12°C over a period of 14 h (between 16:00 to 6:00), savings in the heat and fuel consumption amounting to 18.6÷22% were accomplished. It was also noted that the value of savings was relative to the lengths of pauses in the heating and the accepted indoor temperature during such pauses.

4.6. Lighting and Electrical Appliances

By reducing the thermal energy consumption one should simultaneously reduce the electrical energy consumption. Electrical appliances include lighting, refrigerator, air-conditioning, dish washer, washing machine, iron, television, computer and other appliances, such as hair dryers, and mechanical drives. In the residential sector, electricity is primarily consumed for refrigeration, lighting and air conditioning. Television and computer use 6÷7% of all electrical in the home [171]. Annual electricity consumption of television has increased compared to previous years as a result of an increase in the number of TV channels, a longer daily broadcast period and utilization of color TVs. Energy and exergy efficiencies are assumed to be 75% [172].

A large amount of energy is also used for lighting, which should be also more efficient. In the study [173] it was proposed to reduce wattage by retrofitting incandescent lamps with more efficient compact fluorescent lamps (CFL) in a residential sector in Malaysia. The authors concluded that the lighting retrofit would make a significant impact on residential electricity consumption at the national level and would be a great opportunity for improving the efficiency of the residential sector. This can be achieved only by encouraging consumers to buy and use more efficient appliances through implementing a minimum energy efficiency standard for these appliances. Such programs are now becoming mandatory in many countries, for example in China, Malaysia, Canada and Australia [174-177].

Under a minimum energy efficiency standard, appliances must meet a specified energy efficiency level before they can be legally sold. This is very good solution, because in the past two decades, standards have significantly raised the level of energy efficiency for new products [178].

In the study [179] it was mentioned that the largest energy savings are associated with standards for residential electronics products, but this estimate is subject to a high degree of uncertainty.

Also standards for currently unregulated products may yield more benefits than upgrading minimum efficiency standards for products that already have them.

On the other hand nowadays, because of constantly growing energy consumption and price, more and more often special attention is placed on building design which makes day lighting possible. Natural light (daylight) is a free and simultaneously very effective source of lighting. Its presence in the interior is also an important factor influencing the human psyche. Being present in rooms lit with natural light improves the users frame of mind and their job performance. Utilization of natural light can reduce the electrical power requirement and costs of the air-conditioning and heating, therefore utilization of natural light must become a component of modern building energy management.

Installations of natural lighting are dependent on the location of the object and current meteorological conditions. Inadequate design can cause overheating of rooms during summer. Hence, control systems should be an always inseparable element of natural lighting management systems. The control system should detect the presence of users in the room and switch on necessary or to switch off needless additional artificial lighting.

Chapter 5

TECHNICAL SYSTEMS IN PASSIVE BUILDINGS

Interest in energy-saving building and its practical application appeared together with the change of energy policy and with activities aimed at energy savings and the improvement of the natural environment. The design of ecological and energy-saving buildings is a complicated process, requiring careful cooperation between architectural-planning requirements and technological and ecological requirements. Special project criteria for low energy buildings need to be established for the diminution of energy demand, required to be delivered to the building [180].

The main characteristics of energy-saving buildings are:

- location of the building and special solutions predominantly favourable to the thermal protection of the building,
- very high parameters of the thermal insulation of external and internal partitions, i.e. the introduction of energy-saving solutions involving new or novel constructional -materials,
- good solution for the ventilation of rooms (use of the mechanical supply-exhaust ventilation with heat recovery),
- heating installation with very high efficiency and reliability, equipped with automatic control, measuring, regulatory and weather automatic equipment,
- rational solar energy use, as the factor reducing the energy consumption and the amount of uels delivered to the building,
- use of renewable energy sources, for example: wind energy, hydropower, biomass and communal waste, geothermal energy.

A passive house is a building with extremely low energy demand for heating. In such houses, thermal comfort is assured by internal heat sources; by inhabitants, by heat losses of electric devices, by lighting and by heat gains from the Sun, from heat recovery from the ventilation and in individual cases by heating of ventilating air. Heat losses to the environment are minimized; however the heat gains are used most effectively. In a passive building it is not necessary to design a separate heating system to limit heat losses in the winter and an air conditioning system in the summer. Thermal comfort is assured by passive heat sources [180-184].

5.1. Standards of a Passive Building

For comparison in Table 14 basic criteria and standards of typical, low energy and passive buildings are presented.

Table 14. Criteria and standards of residential buildings – comparison of parameters [180,182,185,186]

Requirements	Typical building	Low energy building	Passive building
Maximum heat transfer coefficient of building partition, W/(m^2·K):			
- external wall,	0.30	0.18	0.15
- roof /flat roof,	0.25	0.20	0.10
- window	1.80 / 1.70	1.70	0.80
Energy demand for heating, W/m^2	≤ 100	25÷35	≤ 10
The annual energy demand for heating, kW·h/(m^2·year)	≤ 100	30÷40	≤ 15
Tightness of a building (n_{50}), h^{-1}	≤ 3.0	...	0.2÷0.6

The standards of a passive house can be reached by introduction of certain technical solutions. Special attention should be paid to [183,187,188]:

- suitable thermal isolation of the external partitions (using traditional thermomodernization one can obtain a diminution of heat demand of ca. 35÷40%),
- reduction/removal of occurrence of thermal bridges,
- stuffing of building cover,

- application of special energy saving windows,
- assurance of effective ventilation with heat recovery from ventilating air.

The maximum heat transfer coefficient of the external partitions (walls, the roof, the floor on the ground) of passive buildings is equal to 0.10÷0.15 $W/(m^2 \cdot K)$. The limitation of heat losses by eliminating thermal bridges is a very difficult target. The linear heat transfer coefficient in virtue of the occurrence of a thermal bridge referred to external dimensions should not exceed 0.01 W/(m·K). To recommended constructions of external partitions of passive buildings can be included:

- brick external wall with thermal insulation made by light wet method and a thickness > 25 cm,
- paned wooden elements produced from wooden I-sections filled with thermal insulation, thickness > 30 cm,
- prefabricated elements from gas concrete with external thermal insulation,
- massive wall from wooden balls with external thermal isolation,
- wall with a vacuum-isolation of 2.5 cm thickness.

Internal partitions should be massive, allowing for the accumulation.

In passive buildings special attention should be paid to the construction of windows. Windows are elements of the building by which losses of 18÷30% of supplied energy can be incurred. Suitable location of glazed surfaces, selecting the type and size of their surface can clearly diminish the energy demand for heating of rooms. The proper location and the construction of windows cause the attainment of maximum heat gains from the solar radiation in the winter and the limitation of intensive heating of rooms in summer period. Accordingly, elements of the building should be fitted to allow exposition to the sun, for example by pushing out of part of the roof outside the south wall. Here the concept of the variable angle of incidence of the solar radiation in winter and summer season is used. If the weather moulding is pushed out of the vertical partition, then it limits the solar radiation of the building, especially when the sun is high above the horizon, as is the case in the warm months.

The amount of solar radiation, which passes to the rooms is related to the angle of incidence of the solar beams. At an angle of incidence of the beams of 30°, the glass of windows of thickness 3 mm absorbs 6% of solar radiation energy, 8% is reflected from it, and 86% passes to the room. In passive buildings the coefficient of the total transmission of the solar radiation energy amounts to > 50÷60%. Windows should be situated on the south or south-west side. From an energy point of view, the optimum is when the ratio of the surface of the south-facing windows to the floor surface of passive buildings of massive construction, is in the range 1/8 to 1/10 of the floor surface.

Heat losses by windows may be limited by:

- diminution of window size,
- additional window pane,
- insulation,
- special sorts of glasses and curtains,
- window shutters.

The heat transfer coefficient for windows (including window frames and casings) in the passive building should not exceed 0.8 $W/(m^2 \cdot K)$. Two-gliding windows with a layer reducing the heat flow, filled with argon with a value of heat transfer coefficient in the range 1.2÷1.6 $W/(m^2 \cdot K)$, do not fulfil requirements of passive building. The temperature of the internal surface of the window could in this instance drop below 14.5°C, and will the decrease the thermal comfort in the room. Hence, in passive buildings three-gliding windows with two layers reducing the heat flow, filled with argon or krypton with a heat transfer coefficient 0.6÷0.8 $W/(m^2 \cdot K)$ should be applied. In this case, the temperature of the internal surface of the window is close to the temperature of the air in the room. The heat transfer coefficient of window panes is ca. 0.7 $W/(m^2 \cdot K)$. Transparent surfaces of windows should make possible rational heat gain from the solar radiation energy, this means the heat gains from solar radiation should exceed the heat losses in all operating situations.

The thermal requirements of profiles of window frames in passive buildings are as follows:

- frames from polyurethane foam strengthened with the steel, aluminium or glass fibre profile,
- frame profiles embossed with two or with three airchambers located internal and outside, with the internal strengthening,
- frames filled with polyurethane foam with the wooden, metal or plastic facing,
- wood frames with soft fibreboard,
- frames from polyurethane substratum [189].

Windows in passive building remain unopened, and are sources of daylighting and solar gains. The system of applied sun-devices can be regulated depending on the outside temperature and conditions of sun exposure. Passive heat gains from the Sun should cover about 40% of energy demand [182].

Passive houses must be very tight and thereby have a mechanical ventilation system with a high efficient heat recovery from the exhaust air. The minimum efficiency of the recuperator should be ca. 75%. The power consumption by the recuperator should be $< 0.45\ W{\cdot}h/m^3$.

The tightness of the building is defined by the difference of the pressure between its internal and the external environment equal to 50 Pa. The air stream infiltrating the building should be smaller than 0.6 of internal capacity per hour ($0.2 \div 0.6\ h^{-1}$). The individual stream of external, fresh air flow into the building should amount to a minimum of 30 m^3/h per person [182]. The passive building should be equipped with energy saving devices in the household and lighting.

Summing up, the passive building should be characterized by the following parameters:

- individual air stream on supply, referred to 1 m^2 of the livable surface, resulting from hygienic needs equal to 1 m^3/h,
- adherence in the air heater to the limitation of temperature < 50°C (less than 46°C) [190],
- coefficient of the annual energy demand for heating of building not exceeding 15 $kW{\cdot}h/(m^2{\cdot}a)$,
- coefficient of total primary energy consumption on all energy needs of the building per year amounting to maximally 120 $kW{\cdot}h/(m^2{\cdot}a)$,
- total final energy consumption maximally 42 $kW{\cdot}h/(m^2{\cdot}a)$,
- the degree of the compactness of the building, defined by the shape factor (A/V), should be at least, but better lower than 0.62 m^{-1}.

5.2. Heating and Hot Water Preparation Systems in Passive Buildings

The aim of the heating and ventilating system in a building should be the limitation of primary energy consumption and the maximum use of renewable energy sources, based on selection of a proper heating medium, that is air and water (with possibly low temperature on supply). Internal installations should be characterized by elasticity, low thermal inertia and limitation of heat losses in heating installation.

Passive buildings use modern technologies not only in the area of thermal isolations or energy savings windows, but also in [191]:

- systems of heat recovery from ventilation,
- photovoltaic systems, using the conversion of solar radiation energy to electrical energy,
- systems of automatic regulation and control (domestic automated technology),
- active and passive solar systems (conversion of solar radiation energy to heat),
- integrated mechanical systems,
- heat storage systems [182].

The heat demand on heating of passive building is not large and amounts to no more than 10 W/m^2. The role of heating installation in passive buildings can fulfil the ventilating installation.

Another solution is the designing and realization of a high efficient low temperature heating system with steel radiators or/and with panel radiators (floor, wall). The temperature of the heating medium on supply should be in the range of 20÷50°C [183].

Application of electric heating may be also considered [192].

The heat source in the heating installation may be a gas condensing boiler or low temperature gas boiler with a closed combustion space and low power. Apart from the conventional energy sources, renewable energy sources such as the following can be used [183]:

- energy of the solar radiation in passive systems as natural daylight [187] and as heat energy (heat gains of sun exposure) or in active systems by use of liquid solar collectors and photovoltaic cell,
- wind energy by use of the autonomous wind power plant for production of electrical energy,
- biomass energy via combustion in a boiler, or in closed compact fireplaces or in fireplaces with a water jacket,

- biogas energy from anaerobic fermentation of animal excrement,
- geothermal energy as the bottom heat source of heat pumps,
- waste energy (heat from the ventilation air, sewage), which is recovered by use of recuperators and heat pumps.

Low power fireplaces heating with biomass candiminish the primary energy demand and be advantageously applied in one-family buildings.

To fulfill the building criteria for energy-saving buildings it is possible to use passive solar air heating [193]. By this concept, the "passive solar heating system" one should understand a structural-materials arrangement, consisting of correlative, helioactive elements of the building without separate installation of the collector, storage tank, pipes and mechanical devices, such as the pump or the ventilator. Their functions fulfil windows, glazed buffer spaces, external collector walls, collector-accumulation or massive internal partitions. In the system of direct heat gains the solar radiation arrives at rooms through windows and is absorbed by the surface of the housing and partly accumulated in massive internal partitions. In systems of indirect heat gains the role of the collector fulfils most often the south wall with dark external surface, situated near behind glazing or separating the heated room from the glazed buffer-space in the form of the veranda or the greenhouse.

An example of such partition can be the classical Trombe wall. This is a massive wall without the heat insulation covered by an exterior glazing with an air channel in between. This wall absorbs and stores the solar energy through the glazing [193]. The absorbed energy of the solar radiation is transported inside the material of wall and with a certain delay passes to rooms or levels heat losses by this wall to the environment.

The collector wall is a light, isolated wall with very small heat capacity. In this instance the heat, from the absorbed solar radiation on its external surface is transported by air in the slit or the space between the wall and protective glazing, and then arrives at the room through openings of circulation.

Among the advantages of passive solar heating systems like these is that:

- they make up an integral part of the building and so do not demand additional investment costs,
- they are characterized by a high level of reliability and work without professional supervision,
- they improve the heat balance of the building,
- they are energy-saving and ecological.

However, to the disadvantage of passive solar heating systems may be counted:

- discontinuity of heat delivery to rooms in consideration of flux densities of the solar radiation, which are variable in time,
- lowest efficiency of the system is in the winter period, i.e. in a period of highest heat demand,
- difficulties in the maintenance of constant air temperature in the total heated capacity of the building,
- the necessity of installing in the building a modern heating source of high efficiency and low thermal inertia, equipped with elements to individually or zonally automatize temperature regulation in individual rooms [194].

In a passive house, the domestic hot water turns out to be the major energy consumer with 50÷80% of the total heat demand [183]. To define the heat demand for hot water preparation one assumes the low level of hot water consumption, is then 25 $dm^3/(d \cdot os)$. In this instance, the most often applied heat sources are:

- a bifunctional gas condensing boiler,
- a heat pump, in this first of all the compact device for heating, ventilation and hot water preparation,
- an automatically supplied boiler using renewable energy, for example based on wood pellets,
- electric floor heating.

All of the mentioned heat sources can work together with the preference installation of fluidic solar collectors. The solar thermal system should normally deliver at least 60% of the yearly heat demand for domestic hot water [183].

Chapter 6

THERMAL ENERGY STORAGE

Heat storage can be accomplished in many different ways, in a wide range of source temperatures, and for the wide range of heat capacity. It may be divided into: low temperature, with storage temperatures of no more than 100÷120°C, medium temperature (120÷500°C) and high temperature > 500°C. The heat accumulation can be short (hours, days) or long-term -(months, seasons, years). Most often, storage systems are used based on specific heat, heat of phase transitions, heat of chemical reactions [194-196].

Heat storage using specific heat is the simplest method. It relies on the rise of temperature of the medium filling the battery. The amount of heat stored is related to the heat capacity of the material, its quantities and the temperatures to which it has to be heated. In low temperature heat storage, water is most often used. The specific heat of water is at least twice as high as other factors applied for this purpose. For storing heat in the range from 120ºC to 500ºC low boiling oils are used which at these temperatures have a low boiling pressure. Above the boiling temperature of oils, liquid metals are applied. A disadvantage of the use of inorganic solids is their low specific heat and thermal conductivity. For these reasons special constructional solutions are necessary. The surface of heat exchange must be very large. Inorganic solids make it possible to realize simple and cheap heat storage often with many years of failure free operation. Characteristic properties of solids applied to heat accumulation are presented in Table 15.

As heat storing media can also be applied: melted waxes, sandstone, concrete, brick, cast iron, quartz, glass. The growth in the amount of stored heat is connected with the linear temperature rise of the medium present in the energy store. This leads to an increase of heat losses to the environment and, in case of constant temperature of the source, decreases the intensity of heat exchange and increases the downtime of the system [194,196,198].

Table 15. Physical properties of inorganic solids used for accumulation of heat [197]

Material	Melting point [°C]	Density [g/cm³]	Specific heat (0÷700°C) [kJ/(kg·K)]
Al_2O_3	2015	3.97	0.8÷1.2
MgO	2800	3.58	0.9÷1.3
SiO_2	1728	2.65	0.8÷1.0
Granite	1200	2.70	0.8÷1.0
Fe	1535	7.90	0.45÷0.8

In heat storage systems based on phase transitions, it is very important to know the enthalpy of phase transition of the material. Phase transitions of the type melting-coagulation, evaporation-condensation occur endo- and exothermically, by small changes in temperature of the deposit with the filling from phasic-variable materials PCM (Phase Changing Materials). One disadvantage of such solutions of heat storage is the lability of the thermal-chemical properties of material in time. Systems using phase transition evaporation-condensation are impractical because of large volume changes in the system during the phase transition. Among phasic-variable materials are usually counted: hydrates, hydrated salts, saturated hydrocarbons, carboxylic acids, esters of higher carboxylic acids, polymers [196].

An advantage of batteries using phase transition heat is the absorption and the return of heat in a very narrow temperature range. The heat storage of phase transitions does not cause the fall of the deposit's exergy, being a result of the equation of the temperature of the deposit during heat accumulation. This results from the fact, that the main amount of heat is received by the storage medium at constant temperature and is used on melting the medium. Bodies whose fusion heat is to be used to heat storage, should fulfil following conditions:

- heat of melting should be as large as practical with a small change of volume during the change of state,
- they should have their highest thermal conductivity and specific heat at high working temperatures,
- in every state of aggregation present in the given operating temperature range, the substance should be display minimal chemical reactivity. In

this way damage to the installaton over many years of operation can be minimised,

- substances should not cause toxicity, explosion or fire danger, they should be easily accessible, and their costs not be large,
- changes of physical properties of the liquid phase in the range of exploitive changes of temperature should be small [197].

In the case of heat storage by use of chemical transformations, the following types of reactions are frequently employed:

- decay reactions of sulphates and nitrates,
- synthesis of ammonia,
- decay of sulphur trioxide to sulphur dioxide and oxygen,
- decay reactions of fluorides,
- decay reactions of hydrides and hydroxides,
- reactions of hydrogenation and dehydrogenation of organic compounds [196,199].

Examples of reversible reactions in heat accumulating systems are presented in Table 16 [196,199,200].

Table 16. Chemical reactions used in the process of heat storage - examples [196,199,200]

Chemical reaction	T [K]	ΔH^0 [kJ/mol]
$CaSO_4 \leftrightarrow CaO + SO_3$	2145	402
$MgSO_4 \leftrightarrow MgO + SO_3$	1470	280
$CaCO_3 \leftrightarrow CaO + CO_2$	1110	178.3
$2SO_3 \leftrightarrow 2SO_2 + O_2$	1040	198.94
$2Ba(OH)_2 \leftrightarrow 2BaO + 2H_2$	1029	77.3
$KBF_4 \leftrightarrow KF + BF_3$	978	109
$CH_4 + H_2O \leftrightarrow 2CO + 2H_2$	961	205
$Ca(OH)_2 \leftrightarrow CaO + H_20$	752	109.9
$NaH \leftrightarrow Na + ½H_2$	735	88
$C_6H_{12} \leftrightarrow C_6H_6 + 3H_2$	568	206.2
$2NH_3 \leftrightarrow N_2 + 3H_2$	466	92.2

ΔH^0 – change in enthalpy of chemical reactions

Reversible reactions of the type solid-gas can also be considered in which the gaseous component could resublime. Storage systems, using the heat of reversible chemical reactions, usually involve very complicated engineering, are characterized with a lower efficiency and are more expensive [196].

6.1. Storage of Heat Energy in Heating Systems

The introduction of capacitive elements to installations may appear in some cases very profitable with respect to economics and functionality. Heat storage is necessary, when:

- heat production and heat demand do not occur simultaneously,
- the necessity to cover peak heat demands appears,
- some heating devices do not assure the heat delivery instantaneously, in the required quantity and suitable quality.

Energy storage systems should be in heating systems in which a heat source is a boiler using solid fuel (coal, biomass), a fireplace with a water jacket and in installations using solar energy.

Buffer tanks make possible the simultaneous connection of several heat sources and supply both the heating and the hot water preparation installations. They are specially recommended when using renewable energy sources. The buffer tank is a special kind of storage tank containing hot water which can be in addition the hydraulic clutch and separator of gases and the mud trap. The main function of the buffer is storage of surpluses of heat energy. A buffer tank can be found in the following installation systems:

- with heat pump, for the purpose of its economic use, independently of the temporary heat demand,
- with solar energy collectors, for heat storage and use of hot water in the night or by appearance of cloudiness,
- with the boiler using solid fuel, with gas or oil boiler,
- in heating and cooling installations to cover peak energy demand.

The buffer, designed in the heating system with a heat pump, fulfils the following targets:

- divides volumetric flows of the factor into the heat pump circuit, where the maintenance of the temperature difference between 5÷7°C is required and the heating circuit, where the temperature difference may fluctuate from 7 to 15°C,
- allows buffering of energy in periods of occurrence of a second tariff of charges for electrical energy,
- eliminates frequent switching of the heat pump in times of low energy demand, increasing the durability of all its elements,
- protects the heat pump against the high temperature of the returning heating factor.

An installation without a buffer tank can take place exclusively in the case of supplyimg only one heating circuit, best in the form of floor panel heating. The large water capacity of the installation will protect the heat pump from large oscillations of temperatures on return. The presence of a sluice valve is then necessary in the installation for the purpose of assuring the volumetric flow on the heating side, which is required by the heat pump. At the design stage, as a rule one should assume the coefficient of selection of ca. 20÷25 (30) dm^3 of buffer tank capacities on 1 kW of the heating power of heat pump.

In a heating installation using a biomass boiler, an accumulation tank of suitable capacity (approximately 40÷60 dm^3 of buffer per kW of boiler power) should be applied to protect against overheating. If, inside the buffer, the heat exchanger (in the form of a pipe coil) can be placed, then the possibility exists of using another alternative heat source in the form of a heat pump or solar collectors. The biomass boiler does not react quickly to changes in heat demand. Boilers using solid fuel or biomass working with nominal power, realize higher efficiencies and cause lower emissions of pollution to the atmosphere as a result of the fuel combustion. The surplus of heat energy produced in the boiler is stored in the tank, and then used, when the boiler does not work at all or works with lower efficiency. The use of an accumulation tank allows a diminution of fuel consumption by as much as 20÷30%.

A buffer heating medium tank can be also in the heating system with Logotherms. Its use is necessary, when the capacity of the supply part of the central heating installation is too small. If the capacity of the supply part of a central heating installation exceeds 300 dm^3, then the buffer tank is not required. The task of the buffer tank is:

- to cover the maximum demand on heating factor in winter period,
- cover of the inequality of draw-offs of heating factor in summer period,

- to assure the necessary amount of heating factor for the correct working of the Logotherm in times of renewed starting of the heat source.

The buffer tank capacity in the system with Logotherms may be defined on the basis of Table 17.

Table 17. Selection of buffer tank capacities in an installation with Logotherms [201]

Number of dwellings	Power of heat source [kW]	Buffer tank of water heating	Tank capacity [dm^3]
1÷10	< 80	Required	330
10÷20	80÷120	Required	260
20÷30	120÷145	According to calculations	150
30÷40	145÷180	According to calculations	150
> 40	> 180	Not required	-

The use of a buffer tank is recommended for the assurance of failure-free functioning of energy systems.

The rule of operation of storage tank based on the cyclic work, that is its download and discharge. On discharge, the heat battery takes over the function of the device supplying the heat. The time of discharge of the storage tank is relative to its volume, the capacity of the building being heated, to the heat demand from individual tasks and to the outdoor temperature [202].

Heat storage makes possible:

- an increase in the efficiency of the energy system,
- an increase in the heat storage in the system and thus an improvement in the degree of its use,
- a diminution of the quantity of fuel, which is combustion,
- a diminution of humidity and the quantity of tar arising during the combustion process,
- a diminution of losses by fractional and incomplete combustion, resulting in reduced pollutant emission,
- improved maintenance of the boiler thanks to the extension of periods between loadings of fuel,
- an extension of boiler exploitation, resulting from the exploitation of the device in conditions of balanced thermal loads [202].

Heat storage can also be implemented in district heating systems. The ability to accumulate heat in the district heating network and in heated buildings during the heating season can smooth daily oscillations of thermal needs. Operating problems can appear resulting in frequent load variations of thermal boilers. This has a disadvantageous influence on the efficiency, reliability of production and heat delivery. Cogeneration blocks reach the highest efficiency with a uniform and nominal load. This is the reason for introducing the heat store, as an additional capacitive element in the district heating network. The use of heat storage is essential both in the process of modernization of existing heat sources, as well as in the construction of new heat sources. A basic task of the heat store is to level the heat demand of the heat source during variable power requirements by recipients, thus assuring stable working conditions [203].

For heating these systems, additional water taken from the local heat system may be used in solar installations with a water storage tank. Under average conditions of sun exposure, the temperature of the additional water should attain 70°C [204].

6.2. Storage of Heat Energy In Solar Systems

Thermal energy gained from solar collectors can be stored in the form of the increase of internal energy (enthalpy) liquid or solids. The choice of kind of the storage medium and the construction of the tank is relative to the type of solar installation and the temperature of heating medium which has to be delivered to the receiver. In solar installations using fluidic solar collectors it is right to use a water storage tank. However, in installations in which in the collector circuit uses air, stony deposit or capsules containing substance(s) with phase transitions may be used as the heat store [146,205].

The energy store increases the efficiency of its processing. When solving a problem of solar energy storage, the following aspects need to be considered:

- the periodicity of its delivery from renewable energy sources,
- power requirement, which is variable in time,
- variety of forms of energy produced and used by recipient,
- no access to other energy sources,
- necessity of effective energy processing,
- ecological issues (for example use of renewable energy sources and waste-energy, diminution of consumption of organic fuels) [194].

In the heat storage process, the basic parameters are: density of storage energy, download and discharge power, the time of download and discharge of the store, the total capacity of the store, the number of download and the discharge cycles, which it is possible to carry out, the operating temperature of the storage system and the total efficiency. The ideal storage system should be characterized by:

- high density of storage energy,
- ease of download and discharge,
- possibility of the simple conversion betweem energy forms,
- small losses in cycles of download and discharge,
- efficiency from the economic point of view and environmentally safe exploitation [194].

Short-term storage is applied in small solar collector systems (to 120 m^2), in systems for hot water preparation, drying in agriculture, and also in heating and air-conditioning. Liquids in these systems (most often water, but also artificial non-freezing liquids) and deposits (natural such as: stones, rocks and artificial structures for example concrete walls, foundations) are used as the storing medium.

Conditions of heat exchange in systems for short-term storage must be chosen so that the system may be able „to be loaded" during source operation. For solar systems this means 6÷8 hours, for systems using the off-peak energy it will be ca. 8 hours [195].

In some systems for heating buildings, thermal energy PCM materials (Phase Changing Materials) are applied for storage. In systems using the specific heat and heat of phase transitions to accumulate heat energy different geometries of deposits and different tanks are required. For PCM materials typical shapes are plates, cylindrical tanks (vertical or horizontal) and spheres. To improve the heat exchange conditions ribbed tanks, modification of the tank's form or application of thermal tubes are used [195].

In systems for the long-term storage of low-temperature energy, the specific heat is mainly used. Accordingly, ground reservoirs filled with water or with water with natural deposit, shallow or deep ground and rocky stores, stores in aquifers, underground excavations or caves filled with water, are applied [195]. The long-term storage can be applied to heating and cooling of buildings and coolings in industrial processes. A typical use of low-temperature storage (i.e. < 100°C) in this context is with natural factors and a tank.

Recommended ranges of operating temperatures of storage systems for energy, stemming from the conversion of solar radiation, are the following:

- for cooling and air-conditioning of rooms: 5°÷17°C,
- for forced heating of flats, offices, shops with warm air: 22°÷30°C,
- for hot water preparation in residential buildings: 47°÷67°C,
- for production and food processing: 47°÷117°C,
- for absorptive cooling: 87°÷180°C [194].

For heat storage inorganic salts, such as: fluorides, chlorides, carbonates, nitrates, phosphates, sulphur compounds and also hydroxides (for example the hydroxide of sodium), oxides (for example boron oxide) and metals can be used.

The energy storage gained from solar collectors by use of phase transitions is considered by many experts as the most interesting practically. At the phase transition considerable amounts of heat, can be absorbed or emitted by small changes in the temperature of the medium. This allows optimal energy use to be delivered by the solar collector system and by optimal selection of the operating temperature of storage to the operating temperature of the source. Heat energy from the thermal conversion of solar radiation can also be effectively stored in reversible chemical reactions or by fuel production (e.g. hydrogen) [194].

6.3. Hot Water Storage Tanks

After the solar collector the storage tank is the second most important element of the installation, influencing the efficiency and operating characteristicses of the system. In solar installations for hot water preparation, a water accumulation tank (the short-term store) is most often used. In this instance in selecting the heat store the following requirements should be considered:

1. The size, construction of tank should be strictly related to the solar collectors surface and adapted to the individual needs of the user [194].
2. Inside the tank, zones with differentiated water temperatures should be allowed to form.
3. During periods of low consumption of hot water, the tank capacity should make possible the storage of accumulated energy [146].

Buffer tanks in solar systems supply the heating and the hot water installations, increasing the energy efficiency of the installation. In a bi-functional solar installation a combi-heater should be included. A combi-heater is a device combining, in one casing, the heat buffer and the heater of the hot water or the buffer preliminary tank. A justification for such a solution is the increased demand on hot water or enlarged partitions of heating medium by additional heating circuits. The buffer tank in solar systems assures preferential heating of hot water and supports the central heating. Recommendations for the selection of a buffer tank for use in solar systems with a low or average degree of participation of solar power are the following:

- in case of hot water preparation one should assume the surface of flat solar collectors to be 1÷1.5 m^2 per person, the surface of vacuum solar collectors 0.6÷0.8 m^2 per person, and the capacity of the water heater according to the coefficient 60÷80 dm^3 per 1 m^2 of solar collector surface,

- for supporting the central heating installation and hot water preparation one should assume the surface of the solar collector to be 0.1÷0.3 m^2 per m^2 of living area, and the capacity of additional buffer tank 60÷80 dm^3 per m^2 of solar collector surface [194].

Water reservoirs used for storage can be directly connencted to the collector circuit or separated from it by means of heat exchangers. This heat exchanger can be situated inside the tank or comprise an external element.

Accumulation tanks in solar installations need to fulfil many requirements. They should be tight, thermally insulated, resistant to hyperbarism and corrosion. In a tank with well installed insulation (thickness 55÷75 mm of rock wool or polymer foam; with the capacity of 200 dm^3 and with a hot water temperature of ca. 60°C), the heat losses to the environment should be within the range of 1.2÷1.5 kWh. If the isolation of the tank is leaky or wet, the daily heat losses can reach even 2.5 kWh. Usually steel tanks are employed, more seldom plastics or concrete. Commercial tanks are covered inside with a special enamel. Important is also the shape of the tank, especially if the phenomenon of thermal stratification is used. With the latter, higher hot water temperatures and better working parameters of the solar collectors can be assured [151].

6.4. STRATIFICATION IN STORAGE TANKS

With regard to the temperature distribution in the tank a tank with full mixing may be considered separately. In this instance the temperature of water in the tank is approximately equal throughout the whole volume. In reservoirs with thermal stratification a distinct vertical gradient of temperatures appears, and water with the highest temperature accumulates in the upper regions of the tank. Water with the lowest temperature is found in the bottom parts of the tank [206]. The differentiation of temperature throughout the tank makes it possible to achieve a higher efficiency (5÷20%) of solar installation for hot water preparation in comparison to an installation in which water in the tank has an equal temperature [207,208]. A stratified tank has much higher performance than a fully mixed the tank [209]. However, thermal destratification caused by mixing of water during its draw-offs, causes an annual decrease in energy efficiency of the system of > 23% [210]. Thermal stratification in the hot water storage tank is very important for the thermal performance of solar domestic hot water systems [146,206,207,209,211-216].

The influence of flow rate and degree of stratification of the water tank on the performance of a classical solar water heating installation has been studied by Cristofari et al [209]. The object of the analysis was a small solar domestic hot water installation for three inhabitants, with a collector area of 2 m^2 and a storage tank of 150 liters. It was shown that using low flow operation results in an increased outlet temperature from the solar collector and a higher degree of thermal stratification in the heat storage. It was established, that for a stratificated tank, an optimal collector flow rate of $2.65 \cdot 10^{-3}$ $kg/(m^2 \cdot s)$ maximizes energy saving and the gained energy is 5.25% higher than a fully mixed tank. For a fully mixed tank, a collector mass flow rate of $30 \cdot 10^{-3}$ $kg/(m^2 \cdot s)$ was assumed.

For tanks directly included in the circuit of solar collector a certain optimum thermal stratification of water exists for which the solar installation under operational and climatic conditions achieves the maximum daily efficiency. The low flow of water in the circuit of the collector enlarges thermal stratification but, at the same time, diminishes the efficiency of the solar collector. However, large values of flow through the collector increase its efficiency, which, unfortunately, at the same time produce strong mixing in the tank. The temperature of water returning to the solar collector is then higher, and the temperature of delivered water to users is lower [146]. In the study [217] Kleinbach et al. ascertained, on the basis of experimental data that, by flow of water transfluent through the tank in a total exchange time shorter than two hours, the thermal stratification practically does not appear.

When the flow rate in the collector circuit is small and when the temperature difference between the hot water inflowing to the tank and the temperature of water inside the tank is large, two zones arise in the tank – an upper zone with high temperature and a bottom zone with low temperature. They are divided by a transitory zone. The phenomenon of thermocline is used to evaluate the quality of heat storage during the download of the tank. For a storage tank loaded with water, flow constant in time and with constant temperature, the energy stored by total loading of the tank can be determined by use of Equation (40) [146,218]:

$$E = V_s \cdot \rho_w \cdot c_w \cdot (T_2 - T_1) \text{ [J]} \tag{40}$$

where:

E – stored energy during total loading of the tank, J,
V_s – total capacity of the tank, m^3,
ρ_w – density of water, kg/m^3,
c_w – specific heat of water inside the tank, J/(kg·K),
T_2 – final temperature of water inside the tank, K,
T_1 – initial temperature of water inside the tank, K.

By slow download without mixing, with ideal thermal stratification in the form of thermocline by the temperature in upper zone (T_2) and bottom (T_1), changes in time of stored energy may be presented in form of the Equation (41) [146]:

$$E_{str} = \dot{m} \cdot c_w \cdot (T_2 - T_1) \cdot \tau \text{ [J]} \tag{41}$$

where:

E_{str} – energy for tank without mixing, J,
$\dot{m}$ - mass stream in collector circuit, kg/s,

τ - loading time [s], variable in the range $0 \le \tau \le \dfrac{V_s \cdot \rho_w}{\dot{m}}$

The degree of loading of the tank is a time function, described by the Equation (42):

$$\frac{E_{str}}{E} = \tau \frac{\dot{m}}{V_s \cdot \rho_w} \tag{42}$$

For the tank with full mixing the change of stored energy may be expressed by use of Equations (43):

$$E_{mix} = \dot{m} \cdot c_w \cdot (T_2 - T_1)[1 - \exp(-\tau \frac{\dot{m}}{V_s \cdot \rho_w})] \text{ [J]} \tag{43}$$

where:

E_{mix} – energy for tank with full mixing, J.

In this case the degree of tank loading can be described using the dependence:

$$\frac{E_{mix}}{E} = 1 - \exp(-\tau \frac{\dot{m}}{V_s \cdot \rho_w}) \tag{44}$$

Thermal stratification inside the tank depends first of all on the volume and shape of the tank, on the location of supply and outlet connections, on the manner of distributing the entering water and on correct location inside the tank of elements potentially able to disturb thermal stratification (heat exchangers, electric heaters) [213]. The recommended heat exchange area should amount to about 0.45 m^2 per m^2 of surface of solar collector absorber. Therefore, the system with direct flow and with the external heat exchanger is most favourable to the formation of the thermal stratification phenomenon. Most difficult is to obtain stratification inside the tank in which the heat exchanger is located inside the tank. A heat exchanger situated in the bottom part of the tank causes full mixing of water in the zone above it. However, locating the low heat exchanger in the upper part of the reservoir causes a decrease in the active volume of the tank. Consequently this limits the amount of energy it is possible to store, because the bottom part of the reservoir, below the heat exchanger, does not participate in energy storage, it is in the so-called „dead zone". An equally disadvantageous solution is to have the heat exchanger in the form of a mantle located on the outside surface of the tank. The mantle has a greater capacity than the internal heat exchanger in form of a spiral and enlarges the amount of liquid in collector circuit. This constellation however, has a negative effect on the ability to adjust the device since it considerably raises the thermal inertia of the installation. The heat exchanger in the form of the mantle generates greater thermal losses to the environment, the heating medium from solar collector circuit with highest temperature has a direct contact with the isolation of the tank, and returns part of the heat to outside the tank.

The presence of thermal stratification inside the tank also influences the optimal lissomeness of the tank. The recommended relation of height of the tank to its diameter should be in the range between 3:1 to 4:1 [216,219]. The thermal stratification within the tank depends on several factors such as heat conduction in the fluid and in the tank wall, heat losses to the surroundings, mixing during charging and discharging processes, geometry of inlet and outlet ports [216,220].

In a well-insulated tank without elements of large conductivity in its interior, a factor having an essential influence on the maintenance of thermal stratification is the conduction inside the tank. The quick disappearance of the vertical temperature gradient is caused by thick walls made from a good heat conductor. Research showed that, for example inside an aluminium tank, the disappearance of thermal stratification appears in a time about 10 times shorter than in the same reservoir made from glass [206]. In a well-insulated tank with walls made of material with low thermal conductance, by simultaneous lack of water flow by through the tank, the thermal stratification can persist for several days.

Consequently, it is best to make the storage tanks from plastics with low thermal conductance. In a correctly designed solar installation the thermal stratification inside the tank forms automatically, and its lack testifies to improper construction of the tank or badly chosen parameters of operation of installation. This happens in installations with an exuded collector circuit and the heat exchanger located directly inside the tank. The improper location of the heat exchanger or badly chosen heat exchange area is the reason for occurrences of convection currents which level the temperature of water inside the tank.

Thermal stratification can be disturbed also by the improper location of the hot water outlet and cold water inlet or the poor construction of connections which lead heated water from the solar collector. Supplementing the reservoir with water should be by means of the inlet situated possibly near the bottom of the reservoir. However, the hot water outlet should be located in the zone above the highest temperature above the reservoir and always above the inlet of the factor from the solar collector. This is particularly important in installations without pressure, filled and emptied gravitationally, where the draft of water below the inlet from the collector would be able to obstruct circulation in the solar collector loop. The connection of return of heated water should assure its low speed at the inlet to the reservoir. Next to the inlet of heated water, a zone of mixing whose volume depends on the speed of the influencing stream always forms.

Thermal stratification in water storage tanks has been the subject of various experimental and numerical studies [220-222]. Thermal stratification can be reached thanks to many methods, including:

- heating the vertical walls of the tank. This causes flow of heated water to the upper parts of the tank;
- heat exchange between liquid contained in the tank and the heating medium, which flows through the heat exchanger. The heat exchanger should be situated in an adequate position inside or outside the tank;
- direct inflow of the warm medium to the tank at a suitable height [220].

In numerical studies, the effect of the modified charging Richardson number, the Peclet number and the Fourier number on the charging efficiency has been presented by Hahne and Chen in [216]. The charging efficiency is above 97%, if the modified Richardson number is greater than 0.25. The experimental study of Sliwinski et al. (1978) has shown that for the modified Richardson number > 0.244, good temperature distribution inside the tank can be achieved. The effect of the Fourier number on the charging efficiency is relatively small.

On the basis of simulation, the effects of different daily draw-off profiles on the performance of domestic hot-water stores using the TRNSYS simulation, was examined. Six daily profiles of hot water consumption were proposed; three complying with real conditions (realistic daily profiles RDPs). It was shown that a large number of small daily hot water draw-offs caused improved efficiency of heat storage. However, the consumption of larger water volumes, as well as draw-offs of longer duration, caused a decrease in the efficiency of the system. The length of time following the hot water consumption from the tank, also influenced the efficiency of the storage system [223].

Results of experimental and numerical studies of thermal stratification within a tank heated by a mantle exchanger was presented in [220]. The thermal stratification was modelled by using a zonal model and the flow in the mantle was modelled by a mixing coefficient.

Using the software TRNSYS the authors analyzed the impact of the cold water inlet device on the thermal stratification in two tanks of 144 liters and 182 liters for small solar domestic hot water systems. It was shown that the thermal stratification inside these tanks depends differently on the flow rate, the draw-off volume, as well as the initial temperature in the tank [215].

CFD (Computational Fluid Dynamics code) calculations and experimental analysis were used to assess the impact of the inlet design on the flow patterns in the tank was investigated. Numerical studies of three inlets with different inlet flow rates illustrated the varying behaviour of the thermal conditions in a solar storage tank [212].

Experimental research concerning thermal stratification in the water storage tank with the internal heat exchanger have been conducted in Poland (Lublin). Inside the tank, heat gained in the vacuum-solar collector for hot water preparation is stored. Technical parameters of a vacuum solar collector and a water storage tank are presented in Table 18.

Table 18. Technical parameters of the evacuated tube solar collector and hot water storage tank [224

Parameter	Description
Vacuum tube collector type “heat-pipe” SG 1800/24: Dimensions: length×width×depth, mm	2040×1994×157
Gross area of collector, m^2	3.90
Glass thickness of a tube, mm	1.6
Inside/outside diameter of a tube, mm	47/58
Insulation thickness of a manifold , mm	55
Insulation material	mineral wool
Fluid conduits diameter, mm	22
Flow channel material	copper
Covering layers on absorbers	AL-N/AL
Absorber absorption	0.95
Maximum efficiency of the collector related to absorber surface	0.623
Collector tilt, °	38
Orientation	South-East
Water storage tank: Tank material	steel
Volume, dm^3	335
Height/diameter, m	1.73/0.5
Insulation material	mineral wool
Insulation thickness, mm	100
External mantle material	steel
Maximum working pressure, bar	6
Maximum working temperature (water), °C	90

The detailed description of solar installation and the measurement system is presented in [224]. The fluid volume flow rate in the circulation was maintained at a constant level of ca. $5.56 \cdot 10^{-5}$ m^3/s.

The efficiency of energy accumulation (η_E) in the solar hot-water system was based on the Equation (45):

$$\eta_E = \frac{\frac{E_s}{A_{col}}}{\sum_{i=1}^{n}(G_i \cdot t_p)} \tag{45}$$

where:

E_s – daily energy accumulated inside the tank, J

$$E_s = m_w \cdot c_w (T_{tn}^{\max} - T_{tn}^{0}) \qquad [J] \tag{46}$$

m_w – mass of water in the storage tank, kg,
c_w –specific heat of water, J/(kg·K),
T_{tn}^{max} – maximum average water temperature in the storage tank, °C,
T_{tn}^{0}- initial water temperature inside the tank, °C,
A_{col} – collector area, m^2,
G – total solar irradiance, W/m^2,
n- number of measuring periods for given day,
t_p – time period, s.

Solar irradiation for a day (SID) was calculated using Equation (47):

$$\text{SID} = \frac{\sum_{i=1}^{n}(G_i \cdot t_p)}{3600000} \qquad [\text{kWh/m}^2] \tag{47}$$

The range of variability of solar radiation and ambient temperature of the solar collector for the days under consideration is shown in Figure 15.

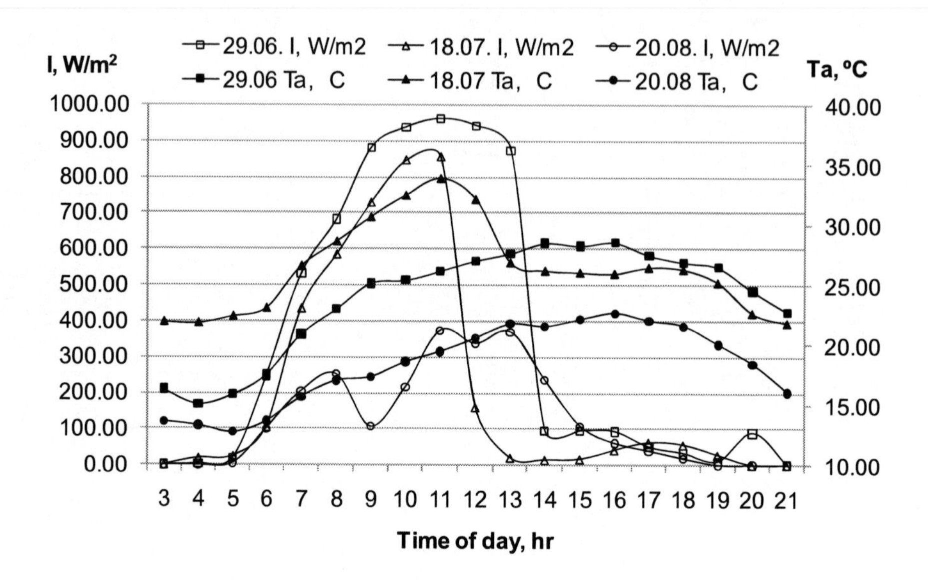

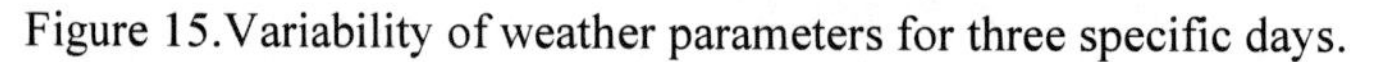
Figure 15.Variability of weather parameters for three specific days.

Figure 16 shows the daily schedules of hot-water consumption on the given days.

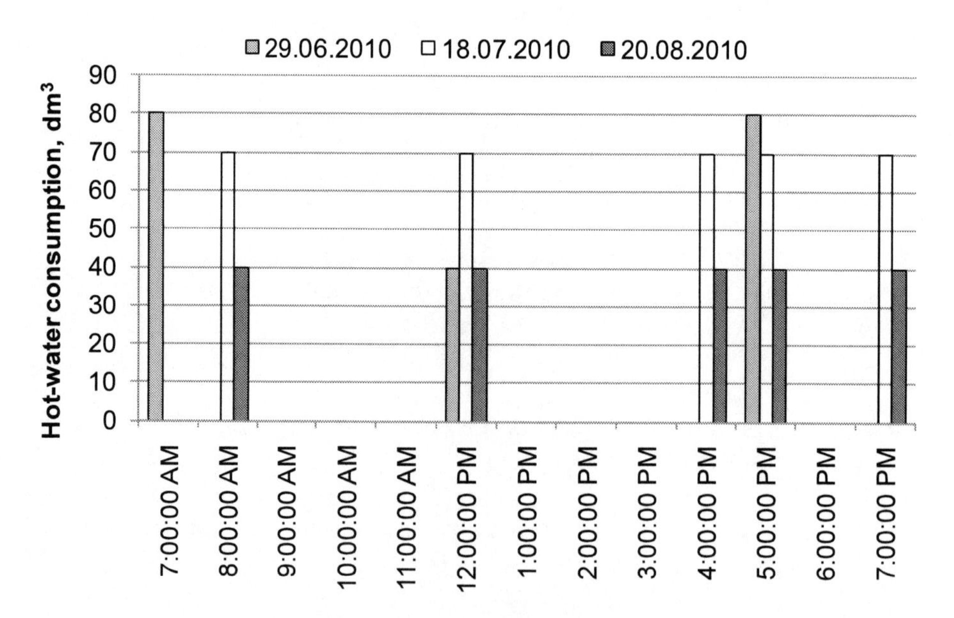

Figure 16. Histograms of daily hot-water consumption for three specific days.

The dynamics of heating (cooling) water at fifteen distinct levels in a storage tank with internal heat exchanger, transferring heat from the solar system for three days in the summer season, are presented in Figures 17, 18, 19. Series T1÷T15 represent the number of temperature sensor. Fifteen sensors were installed every 10 cm looking from the top to the bottom of the tank. For the analysis, all measuring values were assumed as average values within one hour. Figures 20, 21, 22 illustrate the differentiation of temperature inside the storage tank at each level.

Clearly, together with the increase in the daily value of insolation (SID) the energy obtained from solar collectors and the quantity of the energy accumulated inside the storage tank, increased. However, solar irradiation for one day does not have any influence on energy accumulation efficiency in the solar water heating system. The differentiation of water temperature inside the tank appears after loading it in the evening hours and at night. Changes in water temperature are visible when unloading the tank. In most cases during loading the temperature of water in the upper parts of the tank over the pipe coil is fairly uniform.

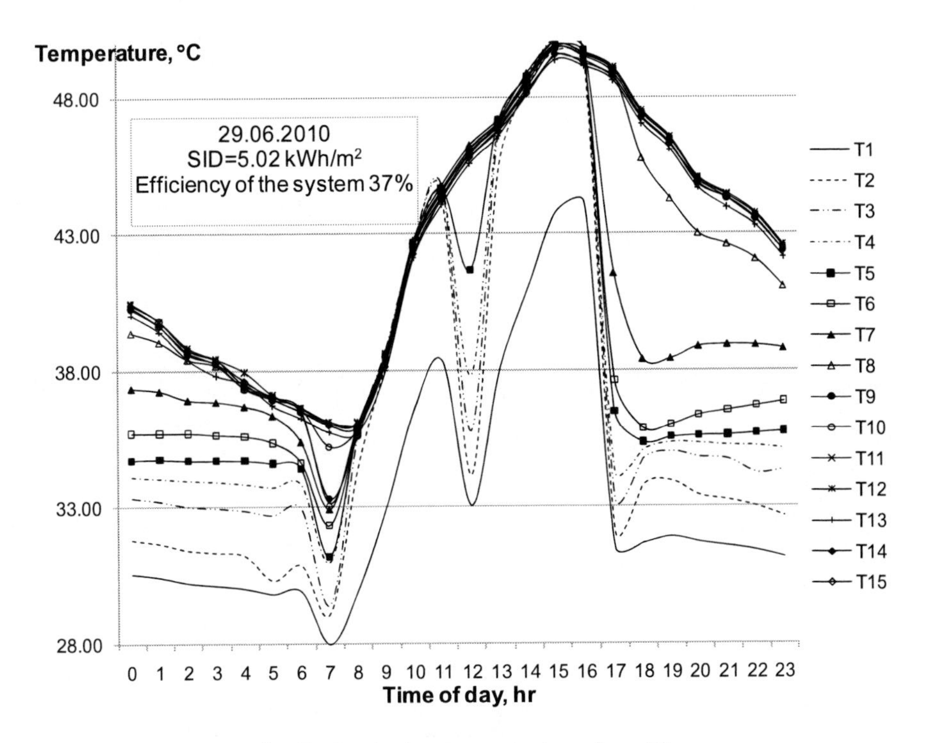

Figure 17. Temperature distribution in the storage tank on June 29.

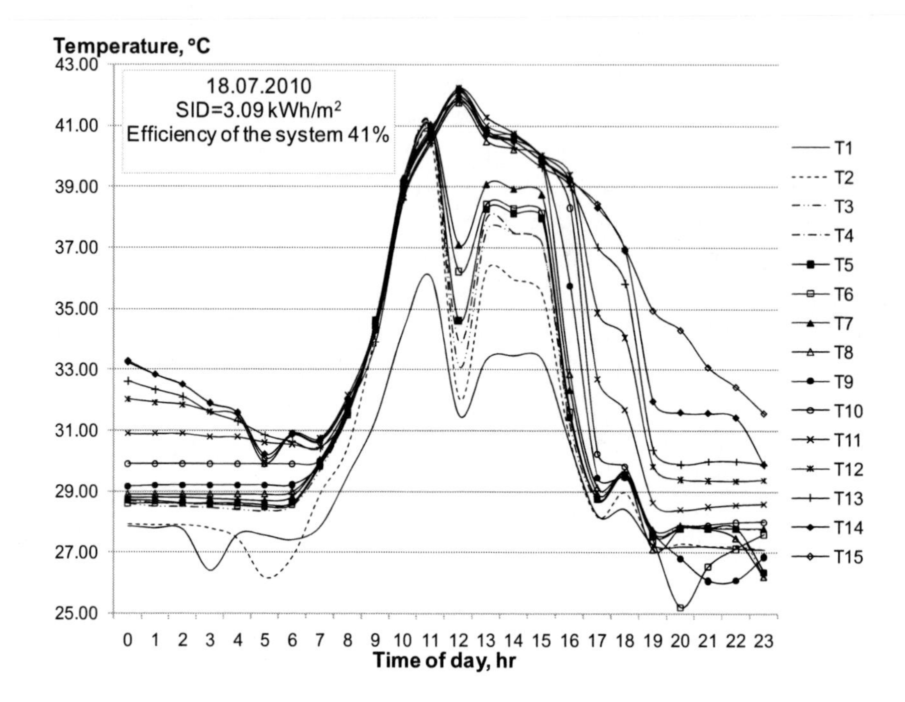

Figure 18. Temperature distribution in the storage tank on July 18.

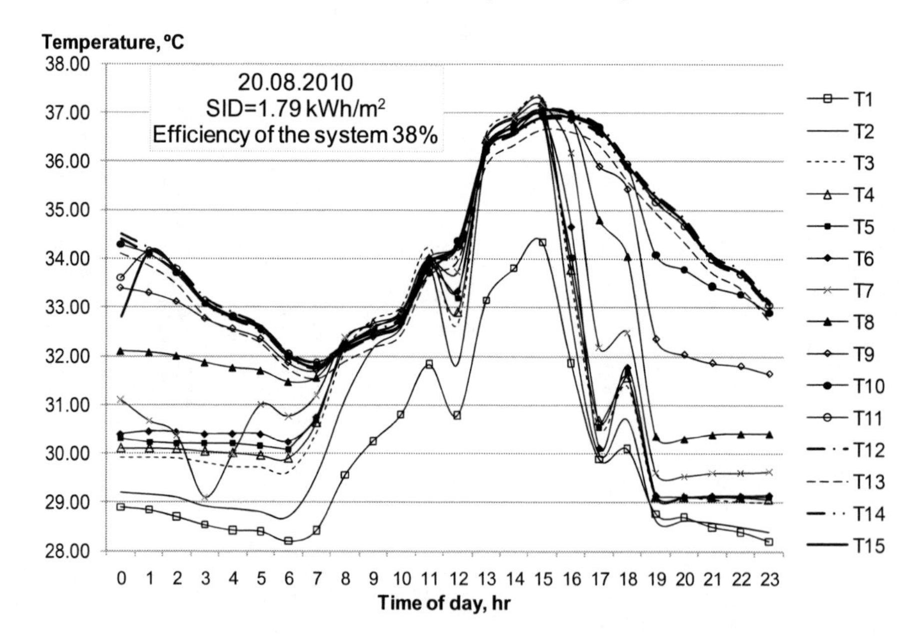

Figure 19. Temperature distribution in the storage tank on August 20.

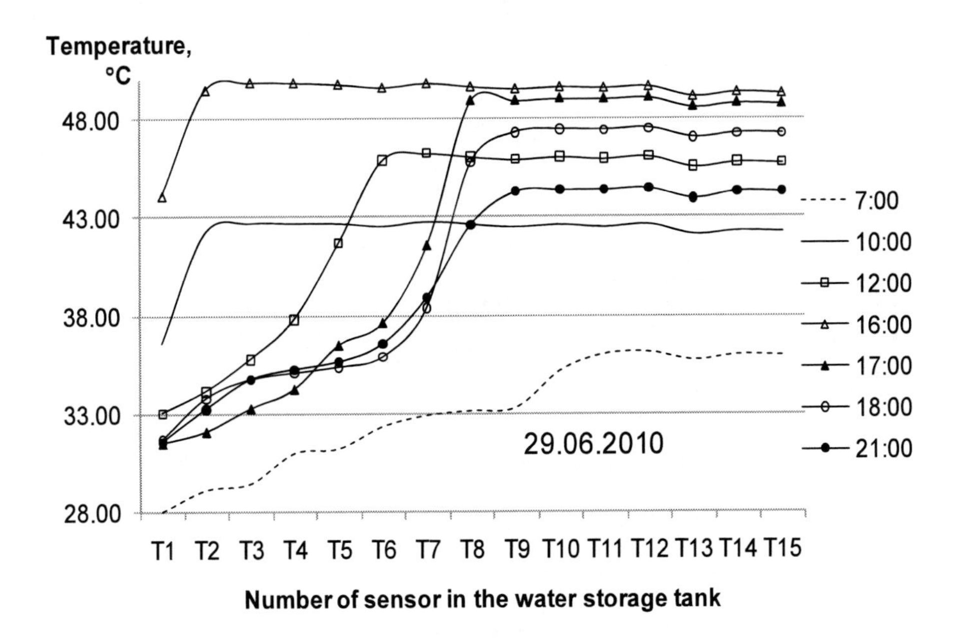

Figure 20. Temperature profile in the storage tank on June 29.

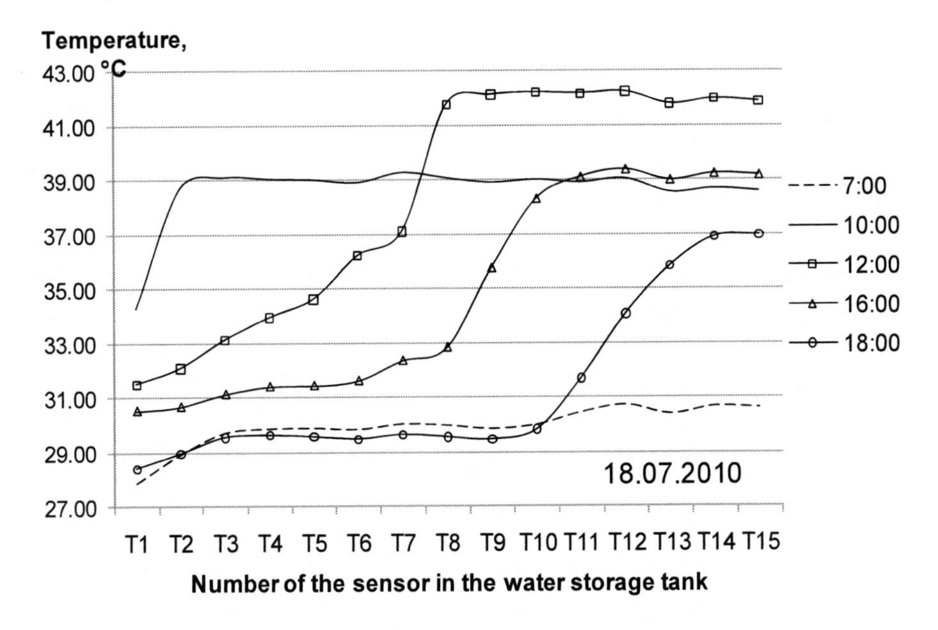

Figure 21. Temperature profile in the storage tank on July 18.

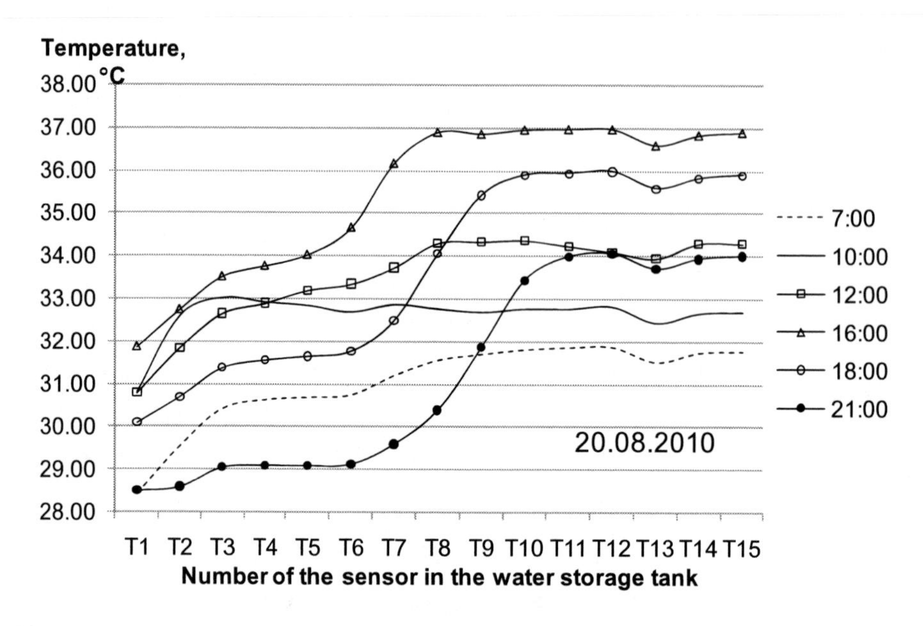

Figure 22. Temperature profile in the storage tank on August 20.

In [225] the simulations on the water temperature distribution inside the storage tank on the basis of TRNSYS software, are presented. For simulation purposes the most suitable TRNSYS type of the tank with stratification, called „Stratified-fluid storage tank" Type 60d, was selected [226]. This type makes possible the simulation of the stratified tank, taking into account the steady-state model of the internal heat exchanger activity. The tank was divided into 15 sections. It was established that:

- there is an optimum-flow in the solar cycle across the internal heat exchanger, which is equal to m=100 kg/h, which improves the thermal stratification level in the solar storage tank. Increasing the optimum-flow to the value of m=350 kg/h causes a slight diminution of the stratification level,
- the change of the heat exchange area of the internal heat exchanger, in the range of 1.0÷2.0 m^2, slightly influences the thermal stratification level in the storage tank,
- the height of the inlet location of the internal heat exchanger influences the thermal stratification level in the storage tank.

An estimation of the energy efficiency of a solar installation for hot water preparation under Polish climatic conditions was carried out using the Exodus method [227]. For the analysis it was assumed that the storage tank was connected directly to the collector circuit, without a heat exchanger. Table 19 presents the technical parameters of this solar collector - water-storage tank system.

Table 19. Data for the flat plate solar collector and hot water storage tank [227]

Parameter	Description
Flat plate solar collector:	
Dimensions: length×width×depth, mm	1920×830×95
Absorber area of collector, m^2	1.20
Glass cover thickness, mm	4
Absorber plate thickness, mm	3
Insulation thickness, mm	50
Flow channel diameter, mm	10
Collective fluid conduits diameter, mm	20
Absorber plate / flow channel material	steel / copper
Covering layers on absorbers	black galvanic chromium on a nickel-plated surface
Insulation material	polyurethane foam + Al film
Glass transmittance	0.80
Absorber plate absorptance	0.95
Maximum efficiency of the collector related to absorber surface	0.79
Collector tilt, °	60
Orientation	South-West
Water storage tank	
Tank material	steel
Volume, dm^3	200
Height/diameter, m	1.5/0.4
Insulation material	foamed polystyrene
sulation thickness, mm	50

A detailed mathematical description of the thermal model of the solar collector making use of the Exodus procedure, as well as the solution of the equation of non-steady heat conduction in the solar collector were presented in [228]. The Exodus procedure was applied in order to define the thermal field in the tank and a detailed description of the application of this calculation method was presented in [227].

Investigations of thermal conditions in the flat-plate solar collector – water-storage tank system enabled the authors to carry out a detailed evaluation of the energy efficiency of the solar installation. Figure 23 presents the average hourly heat efficiencies of the solar collector in comparison to given values of solar radiation intensity and the volumetric flow of water in solar collector circuit for one day.

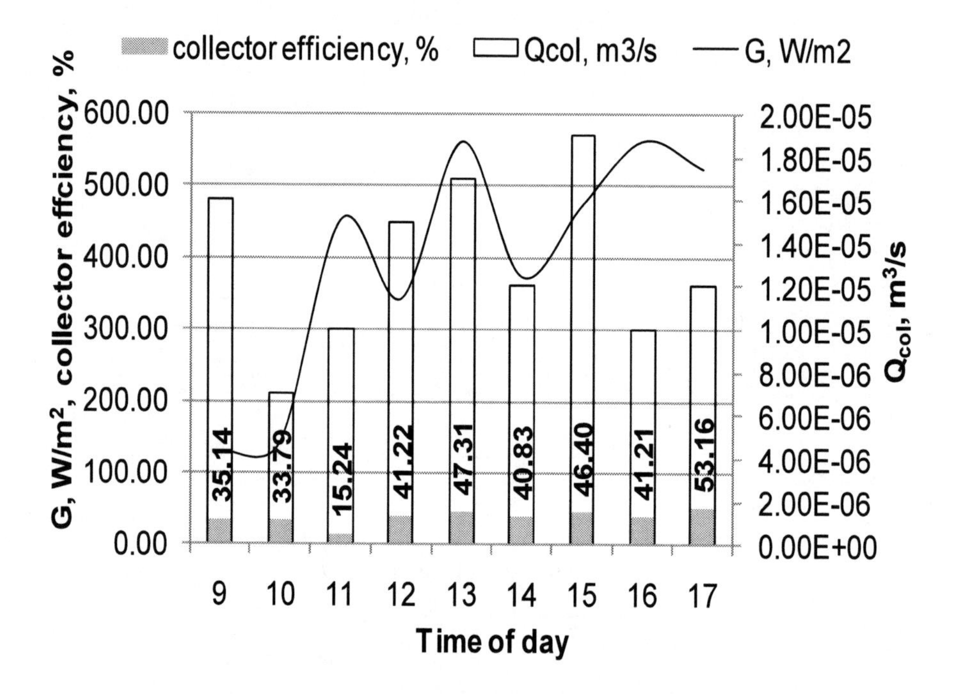

Figure 23. Evolution of the collector's efficiency for real conditions.

The distribution of water temperature for four selected sections inside the storage tank is presented in Figure 24.

For this day at the beginning of the simulation, a homogeneous temperature distribution of water in the storage tank (25.5°C) was assumed. Water in the tank reached the highest temperature of 55.1°C at 5 p.m.

The influence of the flow rate of the medium in the collector cycle, of initial water temperature in the tank and of solar irradiation per day on the quantity of energy accumulated in this system was analyzed. The efficiency of energy accumulation in the tank at the loading stage and during hot water consumption was determined.

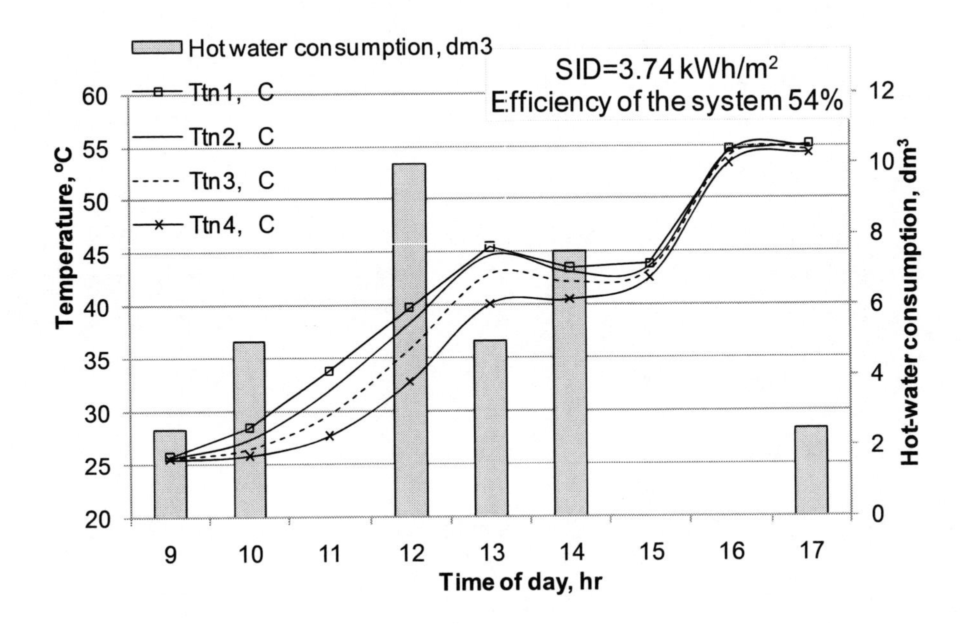

Figure 24. Temperature distribution in the storage tank.

Chapter 7

CONSUMER BEHAVIOUR AND EDUCATION

In addition to the technical possibilities available for energy saving in the residential sector, introduced above, it is crucial to bear in mind the education of end–users [229].

Technical routes cost money and time to bring about improvements in energy efficiency, a change of behavioral patterns by energy-saving education, however can save energy with hardly any additional investment in infrastructure and the energy-saving effect can appear quickly [230,231].

After informing inhabitants about problems connected with the use of fossil fuels and the greenhouse effect as well as indicating the possibility of energy savings in households it is possible to diminish the electrical energy consumption by as much as 10%, as presented in the study [232]. The most important part of energy-saving education is that energy users should have accurate information on which to act.

Further studies [233,234] are available in which all subject households were provided with energy-saving education and tips.

Another method of reducing energy consumption in the residential sector is energy metering [235,236]. Metering gives the user information about the heat consumption and simultaneously stimulates energy-saving behaviour, which can generate energy savings of about 30÷40% [237], or 4÷12% according to [238]. This could generate much energy saving particularly since only 40 percent of those surveyed [239] actually examined their electricity usage when paying the bill.

In the study [240] the authors presented a monitoring system tracking home electricity and natural gas use in heating/cooling and water heating. Sixteen appliances were installed in ten test homes. After completion of the campaign, information on consumption was provided to the residents of the homes. The results indicated that increasing resident awareness of energy use had a major effect on their energy consumption. On average, the total energy use was reduced by 12%.

Acknowledgments

This work was supported by the Ministry of Science and Higher Education of Poland, Grant No. N N523 4136 35.

References

[1] Shafiee, S.; Topal, E. When will fossil fuel reserves be diminished? *Energy Policy,* 2009, 37, 181–189.

[2] PennWell Corporation, Oil & Gas *Journal Energy Database*.

[3] *BP Statistical Review of World Energy*, BP p.l.c.; June 2006.

[4] U.S. Crude Oil, Natural Gas, and Natural Gas Liquids Reserves. *2005 annual report*, DOE/EIA-0216; 2005.

[5] Gulf Publishing Company. World Oil, vol. 227, No. 9; September 2006.

[6] Saidur, R.; Masjuki, H.H.; Jamaluddin, M.Y. An application of energy and exergy analysis in residential sector of Malaysia. *Energy Policy,* 2007, 35, 1050–63.

[7] ACE, COGEN Europe, EuroACE, ECEEE, Eurima, IUT, RICS, 2009. *Open Letter to the 27 EU Energy and Environment Ministers*. Brussels, 2 November 2009.

[8] Itard, L.; Meijer, F.; Vrins, E.; Hoiting, H. Building renovation and modernization in Europe: state of the art review. *OTB, Delft Technical University,* The Netherlands, 2008.

[9] European Commission, Doing more with less, *Green Paper on energy efficiency,* 22.06.2005 COM(2005).

[10] *Energy Efficiency Policies and Measures in Poland, Monitoring of Energy Efficiency in EU 27,* The Polish National Energy Conservation Agency and Central Statistical Office, Warsaw, September 2009.

[11] *www.odyssee-indicators.org*.

[12] Eurostat (*www.ec.europa.eu/eurostat*).

[13] Harrington, L.; Kleverlaan, P. Quantification of residential standby power consumption in Australia: results of recent survey work. *Report for National Appliance and Equipment Energy Efficiency Committee*, Australian Greenhouse Office, Canberra, 2001.

[14] Meier, A.; Lin, J.; Liu, J.; Li, T. Standby power use in Chinese homes. *Energy and Buildings*, 2004, 36, 1211–1216.

[15] COM 2002/91/EC: *Directive on the energy performance of buildings,* EC (2002).
[16] European Commission, *A European strategy for sustainable, competitive and secure energy,* Brussels, COM (2006) 105.
[17] *Kyoto protocol to the United Nations framework convention on climate change,* UNFCCC (1997).
[18] EN, 2007. EN 15217 *Energy Performance of Buildings - Methods for Expressing Energy Performance and for Energy Certification of Buildings.*
[19] SAVE, 1993. *Council Directive 93/76/CEE of 13 September 1993 to limit carbon dioxide emissions by improving energy efficiency* (SAVE).
[20] Perez-Lombard, L.; Ortiz, J.; Gonzalez, R.; Maestre, I.R. A review of benchmarking, rating and labelling concepts within the framework of building energy certification schemes. *Energy and Buildings,* 2009, 41, 272–278.
[21] Swan, L. G.; Ugursal, V. I. Modeling of end-use energy consumption in the residential sector: A review of modeling techniques. *Renewable and Sustainable Energy Reviews,* 2009, 13, 1819–1835.
[22] Seryak, J.; Kissock, K. Occupancy and behavioral affects on residential energy use. *American Solar Energy Society, Proc. Solar Conference,* Austin, Texas, 2003.
[23] Directive 2010/31/EU of the European Parliament and of the Council of 19 May 2010 on the energy performance of buildings.
[24] Polish regulation: Decree of the Ministry of Infrastructure, Dz.U. No. 201, item 1238, 6.11.2008.
[25] Polish regulation: Decree of the Ministry of Infrastructure, Dz.U. No. 201, item 1240, 6.11.2008.
[26] Liddament, M.W.; Orme, M. Energy and ventilation. *Applied Thermal Engineering*, 1998, 18, 1101-1109.
[27] Awbi, H.B. Chapter 7 – Ventilation. *Renewable and Sustainable Energy Reviews*, 1998, 2, 157-188.
[28] Orme, M. Estimates of the energy impact of ventilation and associated financial expenditures. *Energy and Buildings*, 2001, 33, 199-205.
[29] Jokisalo, J.; Kurnitski, J. Performance of EN ISO 13790 utilisation factor heat demand calculation method in a cold climate. *Energy and Buildings*, 2007, 39, 236-247.
[30] Norén, A.; Akander, J.; Isfält, E.; Söderström, O. The effect of thermal inertia on energy requirement in a Swedish building – results obtained with three calculation models. *International Journal of Low Energy and Sustainable Buildings*, 1999, 1.

[31] Catalina, T.; Virgone, J.; Blanco, E. Development and validation of regression models to predict monthly heating demand for residential buildings. *Energy and Buildings*, 2008, 40, 1825-1832.

[32] Jedrzejuk, H.; Marks, W. Optimization of shape and functional structure of buildings as well as heat source utilization. Partial problems solution. *Building and Environment*, 2002, 37, 1037-1043.

[33] Jedrzejuk, H.; Marks, W. Optimization of shape and functional structure of buildings as well as heat source utilization example. *Building and Environment*, 2002, 37, 1249-1253.

[34] Jedrzejuk, H.; Marks, W. Optimization of shape and functional structure of buildings as well as heat source utilization. Basic theory. *Building and Environment*, 2002, 37, 1379-1383.

[35] Depecker, P.; Menezo, C.; Virgone, J.; Lepers, S. Design of buildings shape and energetic consumption. *Building and Environment*, 2001, 36, 627-635.

[36] Jaber, S.; Ajib, S. Optimum, technical and energy efficiency design of residential building in Mediterranean region. *Energy and Buildings*, 2011, 43, 1829-1834.

[37] Park, H.-C.; Kim, H. Heat supply systems using natural gas in the residential sector: The case of the agglomeration of Seoul. *Energy Policy*, 2008, 36, 3843–3853.

[38] Gustafsson, S.I.; Rönnqvist, M. Optimal heating of large block of flats. *Energy and Buildings,* 2008, 40, 1699–1708.

[39] Chłądzyński, S. How to insulate the buildings? *Layers of Roofs and Walls,* 2007, No. 1 (in Polish).

[40] Dylewski, R.; Adamczyk, J. Influence of heating costs on selection of thermal insulation. *District Heating, Heating, Ventilation,* 2008, No. 6, 20-23 (in Polish).

[41] Wichowski, R. The energy consumption in residential buildings in chosen European countries. *District Heating, Heating, Ventilation*, 2007, No. 11, 89-71 (in Polish).

[42] Tommerup, H.; Svendsen, S. Energy savings in Danish residential building stock. *Energy and Buildings,* 2006, 38, 618–626.

[43] Anastaselos, D.; Oxizidis, S.; Papadopoulos, A.M. Energy, environmental and economic optimization of thermal insulation solutions by means of an integrated decision support system. *Energy and Buildings*, 2011, 43, 686–694.

[44] Al-Sanea, S.A.; Zedan M.F. Improving thermal performance of building walls by optimizing insulation layer distribution and thickness for same thermal mass. *Applied Energy*, 2011, 88, 3113–3124.

[45] Duda, L.; Panek, A. The heat market and the support system of the thermal renovation investment. V Conference *„Modernization of District heating systems in Poland”,* Miedzyzdroje, September 1996 (in Polish).
[46] Haralambopoulos, D.A.; Paparsenos, G.F. Assessing the thermal insulation of old buildings – the need for in situ spot measurements of thermal resistance and planar infrared thermography. *Energy Conversion and Management,* 1998, 3, 65-79.
[47] Balaras, C.A.; Arigiriou, A.A. Infrared thermography for building diagnostic. *Energy and Buildings*, 2002, 34, 171-183.
[48] Zalewski, L.; Lassue, S.; Rousse, D.; Boukhalfa, K. Experimental and numerical characterization of thermal bridges in prefabricated building walls. *Energy Conversion and Management*, 2010, 51, 2869–2877.
[49] Duer K.; Svendsen, S.; Mogensen, M. M.; Laustsen, J. B. Energy labelling of glazings and windows in Denmark: calculated and measured values. *Solar Energy,* 2002, 73, 23–31.
[50] Karlsson, J.; Karlsson, B.; Roos, A. A simple model for assessing the energy performance of windows. *Energy and Buildings,* 2001, 33, 641–651.
[51] Solar Energy Laboratory, *University of Wisconsin, TRNSYS reference manual,* 000.
[52] BFRC Guidance Note, www.bfrc.org, 2007.
[53] *Lawrence Berkeley National Laboratory and The University of Minnesota's Center for Sustainable Building Research,* www.efficientwindows.org.
[54] *Energimaerkning Sekretariat Teknologist Institut, Energimaerkning, Energimaerkning,* Tekniske Bestmmelser for vinduer, 2006.
[55] Urbikain, M.K.; Sala J.M. Analysis of different models to estimate energy savings related to windows in residential buildings. *Energy and Buildings,* 2009, 41, 687–695.
[56] Gugliermetti, F.; Bisegna, F. Saving energy in residential buildings: The use of fully reversible windows. *Energy,* 2007, 32, 1235–1247.
[57] Gasparella, A.; Pernigotto, G.; Cappelletti, F.; Romagnoni, P.; Baggio, P. Analysis and modelling of window and glazing systems energy performance for a well insulated residential building. *Energy and Buildings*, 2011, 43, 1030–1037.
[58] Cappelletti, F.; Gasparella, A.; Romagnoni, P.; Baggio P. Analysis of the influence of installation thermal bridges on windows performance: The case of clay block walls. *Energy and Buildings*, 2011, 43, 1435–1442.
[59] Jaworski, J. The energy performance of buildings determined by the thermography. *Polish Solar Energy*, 2005, 3-4, 13-18 (in Polish).

[60] EU Commission, 1997. *A community strategy to promote combined heat and power (CHP) and to dismantle barriers to its development.* COM/97/0514 final EN.

[61] Kleinpeter, M. *Energy Planning and Policy*. John Wiley & Sons, 1995.

[62] Cormio, C.; Dicorato, M.; Minoia, A.; Thorato, M. A regional energy planning methodology including renewable energy sources and environmental constraints. *Renewable and Sustainable Energy Reviews,* 2003, 7, 99–130.

[63] Mróz, T. M. Planning of community heating systems modernization and development. *Applied Thermal Engineering,* 2008, 28, 1844–1852.

[64] Niemyjski, O. Heat losses in district heating networks and possibilities of their limitation. *District Heating, Heating, Ventilation,* 2007, No. 7-8, 9-13 (in Polish).

[65] Stanny, A. The limitation of heat losses with the ConduFill method and their calculation by use of the temperature recorder AS1922G and AS1922T. *District Heating, Heating, Ventilation*, 2007, No. 6, 8-9 (in Polish).

[66] Kamler, W. *District Heating*. PWN, Warsaw 1979 (in Polish).

[67] Wolfe, P. The implications of an increasingly decentralised energy system. *Energy Policy,* 2008, 36, 4509–4513.

[68] Poredos, A.; Kitanovski, A. Exergy loss as a basis for the price of thermal energy. *Energy Conversion and Management,* 2002, 43, 2163–2173.

[69] Wojdyga, K. An influence of weather conditions on heat demand in district heating systems. *Energy and Buildings,* 2008, 40, 2009–2014.

[70] Peeters, L.; Helsen, L.; D'haeseleer, W. IWT-report Development of Extreme Low Energy and Low Pollution buildings by generic optimization: *Control strategies.* K.U. Leuven, Belgium, 2007.

[71] Gram-Hanssen, K.; Bartiaux, F.; Jensen, O.; Cantaert, M. Do homeowners use energy labels? A comparison between Denmark and Belgium. *Energy Policy,* 2007, 35, 2879–2888.

[72] Boryca, J. Economic aspects of the one-family house heating. *District Heating, Heating, Ventilation*, 2008, No. 6, 18-19 (in Polish).

[73] Juanico, L.; Gonzalez, A.D. Thermal efficiency of residential balanced-flue type natural gas space heaters. Measured values for common standard devices. *Energy and Buildings,* 2008, 40, 1067–1073.

[74] Juanico, L. E.; Gonzalez, A. D. Savings on natural gas consumption by doubling thermal efficiencies of balanced-flue space heaters. *Energy and Buildings,* 2008, 40, 1479–1486.

[75] Lee, S.; Kum, S.-M.; Lee, C.-E. An experimental study of a cylindrical multi-hole premixed burner for the development of a condensing gas boiler. *Energy*, 2011, doi:10.1016/j.energy.2011.04.029.

[76] Ghiaus, C.; Belarbi, R.; Allard, F. Optimal settings of residential oil burners. *Energy and Buildings*, 2002, 34, 83–90.

[77] Hanby, V.I. Modeling the performance of condensing boilers. *Journal of the Energy Institute*, 2007, 80, 229-231.

[78] Weiss, M.; Dittmar, L.; Junginger, M.; Patel, M.K.; Blok, K. Market diffusion, technological learning, and cost-benefit dynamics of condensing gas boilers in the Netherlands. *Energy Policy*, 2009, 37, 2962-2976.

[79] Rosiński, M.; Spik, Z. The cost-analysis of the heat consumption on heating and ventilation of one-family buildings depending on applied heat source and the kind of the fuel. *District Heating, Heating, Ventilation*, 2005, No. 3, 23-27 (in Polish).

[80] De Almeida, A.T.; Lopes, A.; Carvalho, A.; Mariano, J.; Nunes, C. Evaluation of fuel-switching opportunities in the residential sector. *Energy and Buildings*, 2004, 36, 195–203.

[81] Shah, V.P.; Debella, D.C.; Ries, R.J. Life cycle assessment of residential heating and cooling systems in four regions in the United States. *Energy and Buildings*, 2008, 40, 503–513.

[82] Alanne, K.; Salo, A.; Saari, A.; Gustafsson, S.-I. Multi-criteria evaluation of residential energy supply systems. *Energy and Buildings*, 2007, 39, 1218–1226.

[83] Gustavsson, L.; Joelsson, A. Life cycle primary energy analysis of residential buildings. *Energy and Buildings*, 2010, 42, 210–220.

[84] Blom, I.; Itard, L.; Meijer, A. LCA-based environmental assessment of the use and maintenance of heating and ventilation systems in Dutch dwellings. *Building and Environment*, 2010, 45, 2362-2372.

[85] Cholewa, T.; Siuta-Olcha, A. Experimental investigations of a decentralized system for heating and hot water generation in a residential building. *Energy and Buildings,* 2010, 42, 183–188.

[86] Lutz, J.D.; Klein, G.; Springer, D.; Howard, B.D. *Residential Hot Water Distribution Systems: Roundtable Session,* Lawrence Berkeley National Laboratory, University of California, 2002.

[87] Hang, Y.; Ying, D.H. Applications of a single house horizontal heating system. In: *Proceedings of the Sixth International Conference for Enhanced Building Operations,* Shenzhen, China, November 6–9, 2006.

[88] Cholewa, T.; Siuta-Olcha, A.; Skwarczyński, M.A. Experimental evaluation of three heating systems commonly used in the residential sector. *Energy and Buildings*, 2011, 43, 2140–2144.

[89] Florides, G.; Kalogirou, S. Ground heat exchangers – a review of systems, models and applications. *Renewable Energy*, 2007, 32, 2461–2478.

[90] Rubik, M. *Heat pumps. Handbook.* Third Edition. Information Centre „Installation Technique in Building Engineering", Warsaw 2006 (in Polish).

[91] Chua, K.J.; Chou, S.K.; Yang, W.M. Advances in heat pump systems: A review. *Applied Energy*, 2010, 87, 3611–3624.

[92] Hepbasli, A.; Kalinci, Y. A review of heat pump water heating systems. *Renewable and Sustainable Energy Reviews*, 2009, 13, 1211–1229.

[93] Rubik, M. Heat pumps – part 4. Lower heat sources – construction and dimensioning of ground heat exchangers. *District Heating, Heating, Ventilation*, 2008, No. 7-8, 6-10 (in Polish).

[94] Milenic, D.; Vasiljevic, P.; Vranjes, A. Criteria for use of groundwater as renewable energy source in geothermal heat pump systems for building heating/cooling purposes. *Energy and Buildings*, 2010, 42, 649-657.

[95] Demir, H.; Mobedi, M.; Ulku, S. A review on adsorption heat pump: problems and solutions. *Renewable and Sustainable Energy Reviews*, 2008, 12, 2381–2403.

[96] Bertsch, S.S.; Groll, E.A. Two-stage air-source heat pump for residential heating and cooling applications in northern US climates. *Int J Refrig*, 2008, 31, 1282–1292.

[97] Chen, L.; Li, J.; Sun, F.; Wu, C. Performance optimization for a two-stage thermoelectric heat-pump with internal and external irreversibilities. *Applied Energy*, 2008, 85, 641–649.

[98] Agrawal, N.; Bhattacharyya, S. Studies on a two-stage transcritical carbon dioxide heat pump cycle with flash intercooling. *Applied Thermal Engineering*, 2007, 27, 299–305.

[99] Wang, W.; Ma, Z.; Jiang, Y.; Yang, Y.; Xu, S.; Yang, Z. Field test investigation of a double-stage coupled heat pumps heating system for cold regions. *Int J Refrig*, 2005, 28, 672–679.

[100] Ma, G.; Li, X. Exergetic optimization of a key design parameter in heat pump systems with economizer coupled with scroll compressor. *Energy Conversion and Management*, 2007, 48, 1150–1159.

[101] Ma, G.Y.; Zhao, H.X. Experimental study of a heat pump system with flash-tank coupled with scroll compressor. *Energy and Buildings*, 2008, 40, 697–701.

[102] Teh, Y.L.; Ooi, K.T. Experimental study of the revolving vane (RV) compressor. *Applied Thermal Engineering*, 2009, 29, 3235–3245.

[103] Teh, Y.L.; Ooi, K.T. Theoretical study of a novel refrigeration compressor – Part II: Performance of a rotating discharge valve in the revolving vane (RV) compressor. *Int J Refrig*, 2009, 32, 1103–1111.

[104] Cascetta, F.; Sasso, M.; Sibilio, S. A metrological analysis of the in situ evaluation of the performance of a gas engine-driven heat pump. *Measurement*, 1995, 16, 209–217.

[105] Colosimo, D.D. Introduction to engine-driven heat pumps—Concepts, approach, and economics. *ASHRAE Transactions*, 1987, 93, 987–996.

[106] Xu, Z.; Yang, Z. Saving energy in the heat-pump air conditioning system driven by gas engine. *Energy and Buildings*, 2009, 41, 206–211.

[107] Kjellsson, E.; Hellström, G.; Perers, B. Optimization of systems with the combination of ground-source heat pump and solar collectors in dwellings. *Energy*, 2010, 35, 2667–2673.

[108] EN 12831:2003 Heating systems in buildings – Method for calculation of the design heat load.

[109] Chudzicki, J. *Installation of domestic hot water in buildings*. Energy Conservation Foundation, Warsaw-Poznan 2006 (in Polish).

[110] Mathys, W.; Stanie, J.; Harmuth, M.; Junge-Mathys, E. Occurrence of Legionella in hot water systems of single-family residences in suburbs of two German cities with special reference to solar and district heating. *International Journal of Hygiene and Environmental Health*, 2008, 211.

[111] Darelid, J.; Lofgren, S.; Malmvall, B.E. Control of nosocomial Legionnaires' disease by keeping the circulating hot water temperature above 55°C: experience from a 10-year surveillance programme in a district general hospital. *J. Hosp. Infect.*, 2002, 50.

[112] Codony, F.; Alvarez, J.; Oliva, J.M.; Ciurana, B.; Company, M.; Camps, N.; Torres, J.; Minguell, S.; Jove, N.; Cirera, E.; Admetlla, T.; Abos, R.; Escofet, A.; Pedrol, A.; Grau, R.; Badosa, I.; Vila, G. Factors promoting colonization by legionellae in residential water distribution systems: an environmental case-control survey. *Eur. J. Clin. Microbiol. Infect. Dis.*, 2002, 21.

[113] International Energy Agency (IEA) Heat Pump Centre; 2009. *<http://www.heatpumpcentre.org>*.

[114] *www.weatheronline.pl.*

[115] Kosieradzki, J. Ground heat exchangers for a heat pump. *District Heating, Heating, Ventilation*, 2008, No. 9, 30 (in Polish).

[116] Blum, P.; Campillo, G.; Munch, W.; Kolbel, T. CO_2 savings of ground source heat pump systems –a regional analysis. *Renewable Energy*, 2010, 35, 122–127.

[117] Lund, J.W.; Freeston, D.H.; Boyd, T.L. Direct application of geothermal energy: 2005 worldwide review. *Geothermics*, 2005, 34, 691–727.

[118] Michopoulos, A.; Bozis, D.; Kikidis, P.; Papakostas, K.; Kyriakis, N.A.Three-years operation experience of a ground source heat pump system in Northern Greece. *Energy and Buildings*, 2007, 39, 328–34.

[119] Rubik, M. Heat pumps – part 5. Lower heat source – construction and dimensioning of ground heat exchangers. *District Heating, Heating, Ventilation*, 2008, No. 9, 3-7 (in Polish).

[120] Reuss, M.; Sanner, B. Auslegung von Wärmequellenanlagen erdgekoppelter Wärmepumpen. *HLH*, 2000, 4.

[121] Mohanraj, M.; Jayaraj, S.; Muraleedharan, C. Performance prediction of a direct expansion solar assisted heat pump using artificial neural networks. *Applied Energy*, 2009, 86, 1442–1449.

[122] Chow, T.T.; Pei, G.; Fong, K.F.; Lin, Z.; Chan, A.L.S.; He, M. Modeling and application of direct-expansion solar-assisted heat pump for water heating in subtropical Hong Kong. *Applied Energy*, 2010, 87, 643–649.

[123] Li, H.; Yang, H. Study on performance of solar assisted air source heat pump systems for hot water production in Hong Kong. *Applied Energy*, 2010, 87, 2818–2825.

[124] Zhai, X.Q.; Qu, M.; Yu, X.; Yang, Y.; Wang, R.Z. A review for the applications and integrated approaches of ground-coupled heat pump systems. *Renewable and Sustainable Energy Reviews*, 2011, 15, 3133–3140.

[125] Olesen, B.W.; Mortensen, E.; Thorshauge, J. Thermal comfort in a room heated by different methods. Technical Paper No. 2256, Los Angeles Meeting, *ASHRAE Transactions* 86, 1980.

[126] Myhren, J.A.; Holmberg, S. Flow patterns and thermal comfort in a room with panel, floor and wall heating. *Energy and Buildings*, 2008, 40, 524–536.

[127] Juusela, M.A. (Ed.) *Heating and Cooling with Focus on Increased Energy Efficiency and Improved Comfort*, Guidebook to IEA ECBCS Annex 37, Low Energy Systems for Heating and Cooling of Buildings. VTT Building and Transport, Espoo, ISBN 951-38-6489-8, 2003.

[128] Hasan, A., Kurnitski, J.; Jokiranta, K. A combined low temperature water heating system consisting of radiators and floor heating. *Energy and Buildings*, 2009, 41, 470–479.

[129] Tsilingiridis, G.; Martinopoulos, G. Thirty years of domestic solar hot water systems use in Greece – energy and environmental benefits – future perspectives. *Renewable Energy*, 2010, 35, 490–497.

[130] Solangi, K.H.; Islam, M.R.; Saidur, R.; Rahim, N.A.; Fayaz, H. A review on global solar energy policy. *Renewable and Sustainable Energy Reviews*, 2011, 15, 2149–2163.

[131] Tsoutsos, T.; Frantzeskaki, N.; Gekas, V. Environmental impacts from the solar energy technologies. *Energy Policy*, 2005, 33, 289–296.

[132] Wang, Q.; Qiu, H.-N. Situation and outlook of solar energy utilization in Tibet, China. *Renewable Sustainable Energy Reviews*, 2009, 13, 2181–2186.

[133] Saidur, R.; Islam, M.R.; Rahim, N.A.; Solangi, K.H. A review on global wind energy policy. *Renewable Sustainable Energy Reviews*, 2010, 14, 1744–1762.

[134] Albers, J.; Dommel, R.; Montaldo-Ventsam, H.; Nedo, H.; Übelacker, E.; Wagner, J. *Central heating and ventilation systems. Guide for designers and installers*. WN-T, Warsaw 2007 (in Polish).

[135] Popiołek, M. Solar collectors as alternative energy source. *Building Materials*, 2008, 425, 75–77 (in Polish).

[136] Choromański, P.; Mikucki, O. Solar energy collectors - heat from renewable energy sources. *Building Materials*, 2008, 430, 77–78 (in Polish).

[137] Jaromi-Wolniakowska, G.; Kochan, K. Solar collectors instead of coal. *Aura*, 2007, 5, 2 (in Polish).

[138] Ickiewicz, I. Economic performance of solar collectors. *Building Materials*, 2007, No. 3, 39,53 (in Polish).

[139] Kasperski, J.; Drzeniecka-Osiadacz, A.; Lewkowicz, M. Simulation of solar radiation intensity on basis of actinometrical measurement data. *District Heating, Heating, Ventilation*, 2007, No. 4, 19-25 (in Polish).

[140] Thirugnanasambandam, M.; Iniyan, S.; Goic, R. A review of solar thermal technologies. *Renewable and Sustainable Energy Reviews*, 2010, 14, 312–322.

[141] Jaisankar, S.; Ananth, J.; Thulasi, S.; Jayasuthakar, S.T.; Sheeba, K.N. A comprehensive review on solar water heaters. *Renewable and Sustainable Energy Reviews*, 2011, 15, 3045–3050.

[142] Adamczyk, J.; Cholewa, T. Experimental investigations of large-scale solar collector installations in an inhabited cloister: A 6-year case study. *Ecological Chemistry and Engineering A*, 2010, 17, 1461-1472.

[143] Pluta, Z. *Solar energy installations*. Publishers of Warsaw University of Technology, Warsaw 2008 (in Polish).

[144] Chang, J.M.; Leu, J.S.; Shen, M.C.; Huang, B.J. A proposed modified efficiency for thermosyphon solar heating systems. *Solar Energy*, 2004, 76, 693–701.

[145] Abreu, S.L.; Colle, S. An experimental study of two-phase closed thermosyphons for compact solar domestic hot-water systems. *Solar Energy*, 2004, 76, 141–145.

[146] Pluta, Z. *Thermal conversion of solar energy: theoretical basis*. First Edition Publishers of Warsaw University of Technology, Warsaw 2000 (in Polish).

[147] Pluta Z. *Thermal conversion of solar energy: theoretical basis*. Second Edition. Publishers of Warsaw University of Technology, Warsaw 2006 (in Polish).

[148] Czekalski, D.; Obstawski, P. Performance of solar heating systems in houses on the basis of the operational research. *District Heating, Heating, Ventilation*, 2008, No. 1, 15-19 (in Polish).

[149] Duffie, J.A.; Beckman, W.A. *Solar engineering of thermal processes*. John Wiley & Sons, New York 1991.

[150] Khalifa, A.J.N.; Jabbar, R.A.A. Conventional versus storage domestic solar hot water systems: A comparative performance study. *Energy Conversion and Management*, 2010, 51, 265–270.

[151] Wiśniewski, G.; Gołębiowski, S.; Gryciuk, M.; Kurowski, M. *Solar collectors. Handbook of use of solar energy*. Third Edition, Publishers: Centre of Buildings Engineering Information, Warsaw 2006 (in Polish).

[152] Hahne, E. Parameter effects on design and performance of flat plate solar collectors. *Solar Energy*, 1985, 34, 497-504.

[153] Bello, M.B.; Sambo, A.S. Simulation studies on pipe spacings for a collector and tank size for solar water heater. *Energy Conversion Management*, 1992, 33, 215-223.

[154] Hegazy, A. Effect of dust accumulation on solar transmittance through glass covers of plate-type collectors. *Renewable Energy*, 2001, 22, 525–540.

[155] Soulayman, S.S.H. On the optimum tilt of solar absorber plates. *Renewable Energy*, 1991, 1, 551-554.

[156] Kasperski, J. Solar collectors inclined sideways and located on roofs having orientation other than south direction. *District Heating, Heating, Ventilation*, 2007, No. 2, 30-33 (in Polish).

[157] Cholewa, T.; Siuta-Olcha, A. Analysis of selected factors affecting energy performance of solar systems for preparation of usable hot water. *District Heating, Heating, Ventilation*, 2010, No. 6, 198-201 (in Polish).

[158] Yohani,s Y.G.; Popel, O.; Frid, S.E.; Norton, B. The annual number of days that solar heated water satisfies a specified demand temperature. *Solar Energy*, 2006, 80, 1021–1030.

[159] Gastli, A.; Charabi, Y. Solar water heating initiative in Oman energy saving and carbon credits. *Renewable and Sustainable Energy Reviews*, 2011, 15, 1851–1856.

[160] Thür, A.; Furbo, S.; Shah, L.J. Energy savings for solar heating systems. *Solar Energy*, 2006, 80, 1463–1474.

[161] Residential Energy Consumption Survey (RECS), Energy Information Administration, US Department of Energy, 2005 *http://www.eia.doe.gov/emeu/recs/contents.html.*

[162] Meyers, R.J.; Williams, E.D.; Matthews, H.S. Scoping the potential of monitoring and control technologies to reduce energy use in homes. *Energy and Buildings,* 2010, 42, 563–569.

[163] Williams, E.; Scott Matthews, H.; Breton, M.; Brady, T. Use of a computer-based system to measure and manage energy use in the home. In: *2006 IEEE International Symposium on Electronics and the Environment, IEEE,* Piscataway, New Jersey, 2006, 161–166.

[164] Peeters, L.; Van der Veken, J.; Hens, H.; Helsen, L.; D'haeseleer, W. Control of heating systems in residential buildings: Current practice. *Energy and Buildings,* 2008, 40, 1446–1455.

[165] Boait, P.J.; Rylatt, R.M. A method for fully automatic operation of domestic heating. *Energy and Buildings,* 2010, 42, 11–16.

[166] Critchley, R.; Gilbertson, J.; Grimsley, M.; Green, G. Living in cold homes after heating improvements: evidence from Warm-Front, England's home energy efficiency scheme. *Applied Energy*, 2006, 84, 147–158.

[167] Ogonowski, S. Modeling of the heating system in small building for control. *Energy and Buildings,* 2010, 42, 1510–1516.

[168] Privara, S.; Siroky, J.; Ferkl, L.; Cigler, J. Model predictive control of a building heating system: The first experience. *Energy and Buildings*, 2011, 43, 564–572.

[169] Cho, S.H.; Zaheer-Uddin, M. Predictive control of intermittently operated radiant floor heating systems. *Energy Conversion and Management*, 2003, 44, 1333–1342.

[170] Szpil, Z. Heat from the environment – Part 3. *Aura*, 2008, No. 4 (in Polish).

[171] Utlu, Z.; Hepbasli, A. Parametrical investigation of the effect of dead (reference) state on energy and exergy utilization efficiencies of residential–commercial sectors: A review and an application. *Renewable and Sustainable Energy Reviews,* 2007, 11, 603–634.

[172] Utlu, Z.; Hepbasli, A. Analysis of energy and exergy use of the Turkish residential–commercial sector. *Building and Environment,* 2005, 40, 641–655.

[173] Mahlia, T.M.I.; Said, M.F.M.; Masjuki, H.H.; Tamjis, M.R. Cost-benefit analysis and emission reduction of lighting retrofits in residential sector. *Energy and Buildings,* 2005, 37, 573–578.

[174] Lu, W. Potential energy savings and environmental impact by implementing energy efficiency standard for household refrigerators in China. *Energy Policy,* 2006, 34, 1583–1589.

[175] Mahlia, T.M.I.; Masjuki, H.H.; Saidur, R.; Choudhury, I.A.; Nooleha, A.R. Projected electricity savings from implementing minimum energy efficiency standards for household refrigerators in Malaysia. *Energy,* 2003, 28, 751–754.

[176] Mahlia, T.M.I.; Masjuki, H.H.; Choudhury, I.A.; Saidur, R. Potential CO_2 reduction by implementing energy efficiency standard for room air conditioner in Malaysia. *Energy Conversion and Management,* 2001, 42, 1673–1685.

[177] Varman, M.; Masjuki, H.H.; Mahlia, T.M.I. Electricity savings from implementation of minimum energy efficiency standard for TVs in Malaysia. *Energy and Buildings,* 2005, 37, 685–689.

[178] Meyers, S.; McMahon, J.; McNeil, M. *Realized and prospective impacts of US energy Efficiency standards for residential appliances.* Lawrence Berkeley National Laboratory. Berkeley, CA. Report No. LBNL-56417, 2005.

[179] Rosenquist, G.; McNeil, M.; Iyer, M.; Meyers, S.; McMahon, J. Energy efficiency standards for equipment: Additional opportunities in the residential and commercial sectors. *Energy Policy,* 2006, 34, 3257–3267.

[180] Audenaert, A.; De Cleyn, S.H.; Vankerckhove, B. Economic analysis of passive houses and low-energy houses compared with standard houses. *Energy Policy*, 2008, 36(1), 47-55.

[181] Piotrowski, R.; Wnuk, R. *A catalog of projects for passive and low-energy buildings*. Publishers: Building Guide Book, Warsaw 2006 (in Polish).

[182] Wnuk, R.: *Installations in a passive and low-energy building*. Publishers: Building Guide Book, Warsaw 2007 (in Polish).

[183] Badescu, V.; Sicre, B. Renewable energy for passive house heating Part I. Building description. *Energy and Buildings*, 2003, 35, 1077-1084.

[184] Persson, M.; Roos, A.; Wall, M. Influence of window size on the energy balance of low energy houses. *Energy and Buildings*, 2006, 38, 181-188.

[185] Mahdavi, A.; Doppelbauer, E. A performance comparison of passive and low-energy buildings. *Energy and Buildings*, 2010, 42, 1314-1319.

[186] Wojdyga, K. An investigation into the heat consumption in a low-energy building. *Renewable Energy*, 2009, 34, 2935-2939.

[187] Llovera, J.; Potau, X.; Medrano, M.; Cabeza, L.F. Design and performance of energy-efficient solar residential house in Andorra. *Applied Energy*, 2011, 88, 1343-1353.

[188] Parker, D.S. Very low energy homes in the United States; perspectives on performance from measured data. *Energy and Buildings*, 2009, 41, 512-520.

[189] Schlagowski, G.; Feist, W.; Munzenberg, U.; Thumulla, J.; Schulze, B. *Handbook of Passive Buildings*. Polish Passive House Institute, Gdansk, 2006 (in Polish).

[190] Badescu, V.; Sicre, B. Renewable energy for passive house heating Part II. Model. *Energy and Buildings*, 2003, 35, 1085-1096.

[191] Chwieduk, D. Towards sustainable-energy buildings. *Applied Energy*, 2003, 76, 211-217.

[192] Feist, W.; Schnieders, J.; Dorer, V.; Haas, A. Re-inventing air heating: convenient and comfortable within the frame of the passive house concept. *Energy and Buildings*, 2005, 37, 1186-1203.

[193] Chan, H.; Riffat S.B.; Zhu, J. Review of passive solar heating and cooling technologies. *Renewable and Sustainable Energy Reviews*. 2010, 14, 781-789.

[194] Gogół, W. et al. *Thermal conversion of solar energy in Polish conditions. Ekspertise*. Committee of Thermodynamics and Combustion, IV Faculty of Technical Sciences Polish Academy Sciences, Warsaw 1993 (in Polish).

[195] Domański, R. *Thermal energy storage*. PWN, Warsaw 1990 (in Polish).

[196] Gut-Wygonik, I. Heat storage. *District Heating, Heating, Ventilation*, 2006, No. 1, 18-19, 22-25 (in Polish).

[197] Kaiser, H. *Use of solar energy*. AGH, Krakow 1995 (in Polish).

[198] Mikielewicz, J.; Cieśliński, J.T. *Unconventional devices and energy conversion systems.* Polish Academy Sciences, Ossolineum, Wroclaw 1999 (in Polish).

[199] Pielichowski, K.; Flejtuch, K. Use of phase changing materials in thermal energy storage. *Fuels and Energy Economy*, 2003, No. 1, 7-12 (in Polish).

[200] Lewandowski, W.M. *Environmentally friendly renewable energy sources*. WN-T, Warsaw 2001 (in Polish).

[201] Residential thermal station. A heating system for a new generation. *Technical information and design. Meibes* (www.meibes.pl) (in Polish).

[202] Zuwała, J. Heating appliances and systems with heat accumulation for individual customer purpose. *District Heating, Heating, Ventilation*, 2007, No. 2, 28-29 (in Polish).

[203] Zwierzchowski, R. Application of heat accumulators in district heating systems. *District Heating, Heating, Ventilation*, 2009, No. 1, 3-6 (in Polish).

[204] Pluta, Z; Wnuk, R. Simple mathematical modeling of solar water heating plant with thermal stratified storage tank. XIII *Symposium on Heat Transfer and Mass*, 2007, 815-822.

[205] Duffie, J.A.; Beckman, W.A. *Solar engineering of thermal processes*. John Wiley&Sons, Hoboken, New Jersey, USA, 2006.

[206] Hollands, K.G; Lightstone, M.F. A review of low-flow, stratified-tank solar water heating systems. *Solar Energy*, 1989, 43, 97-105.

[207] Smolec, W. *Thermal conversion of solar energy*. PWN, Warsaw, 2000 (in Polish).

[208] Jordan, U.; Vajen, K. Influence of the DHW load profile on the fractional energy savings: a case study of solar combi-system with TRNSYS simulations. *Solar Energy*, 2000, 69, 197-208.

[209] Cristofari, C.; Notton, G.; Poggi, P.; Louche, A. Influence of the flow rate and the tank stratification degree on the performances of a solar flat-plate collector. *International Journal of Thermal Sciences*, 2003, 42, 455-469.

[210] Andersen, E.; Furbo, S. Thermal destratification in small standard solar tanks due to mixing during tapping. In *Proceedings of ISES Solar World Congress*, Jerusalem, Israel, 1999, Vol. III, 1197–1206.

[211] Furbo, S.; Andersen, E.; Thür, A.; Shah, L.J.; Andersen, K.D. Performance improvement by discharge from different levels in solar storage tanks. *Solar Energy*, 2005, 79, 431-439.

[212] Shah, L.J.; Furbo, S. Entrance effects in solar storage tanks. *Solar Energy*, 2003, 75, 337-348.

[213] Zachár, A.; Farkas, I.; Szlivka, F. Numerical analyses of the impact of plates for thermal stratification inside a storage tank with upper and lower inlet flows. *Solar Energy*, 2003, 74, 287-302.

[214] Knudsen, S.; Furbo, S. Thermal stratification in vertical mantle heat-exchangers with application to solar domestic hot-water systems. *Applied Energy* 2004, 78, 257-272.

[215] Jordan, U.; Furbo, S. Thermal stratification in small solar domestic storage tanks caused by draw-offs. *Solar Energy*, 2005, 78, 291-300.

[216] Hahne, E.; Chen, Y. Numerical study of flow and heat transfer characteristics in hot water stores. *Solar Energy*, 1998, 64, 9-18.

[217] Kleinbach, E.M.; Beckman, W.A.; Klein, S.A. Performance study of one-dimensional models for stratified thermal storage tanks. *Solar Energy*, 1993, 50, (2), 155-166.

[218] Pluta, Z.; Wnuk, R. Thermal energy storage tanks in solar systems. *District Heating, Heating, Ventilation*, 1997, No. 10, 15-16, 30-32 (in Polish).

[219] Lavan, Z.; Thompson, J. Experimental study of thermally stratified hot water storage tanks. *Solar Energy*, 1977, 19, 519-524.

[220] Kenjo, L.; Inard, C.; Caccavelli, D. Experimental and numerical study of thermal stratification in a mantle tank of a solar domestic hot water system. *Applied Thermal Engineering*, 2007, 27, 1986-1995.

[221] Shin, M.S.; Kim, H.S.; Jang, D.S.; Lee, S.N.; Yoon, H.G. Numerical and experimental study on the design of a stratified thermal storage system. *Applied Thermal Engineering*, 2004, 24, 17-27.

[222] Bouhdjar, A.; Harhad, A. Numerical analysis of transient mixed convection flow in storage tank: influence of fluid properties and aspect ratios on stratification. *Renewable Energy*, 2002, 25, 555-567.

[223] Spur, R.; Fiala, D.; Nevrala, D.; Probert, D. Influence of the domestic hot-water daily draw-off profile on the performance of a hot-water store. *Applied Energy*, 2006, 83, 749-773.

[224] Siuta-Olcha, A.; Cholewa, T. Research of thermal stratification processes in water accumulation tank in exploitive conditions of solar hot water installation. In: *Proceedings of 41 International Congress on Heating, Refrigerating and Air-Conditioning*, Edited by Branislav Todorović, ISBN 978-86-81505-55-7, 391-400.

[225] Siuta-Olcha, A.; Cholewa, T. Simulation research of thermal stratification processes in the water storage tank. In: *Proceedings of the XIII International Symposium on Heat Transfer and Renewable Sources of Energy* HTRSE 2010, Szczecin-Miedzyzdroje, 9-12 September 2010, 259-265.

[226] Klein, S.A.; Beckman, W.A.; Mitchell, J.W.; Duffie, J.A.; Duffie, N.A.; Freeman, T.L. *TRNSYS 15a transient simulation and program.* Madison, Solar Energy Laboratory, University of Wisconsin, 2000.

[227] Siuta-Olcha, A. Energy efficiency evaluation of a solar domestic hot-water system: a case study. *Environment Protection Engineering*, Vol. 36, No. 4, 2010, 23-35.

[228] Siuta-Olcha, A. Application of a probabilistic method for determination of the thermal field in a flat-plate solar collector. *Archives of Environmental Protection*, 2007, Vol. 33, No. 4, 67-81.

[229] Wood, G.; Newborough, M. Dynamic energy-consumption indicators for domestic appliances: environment, behavior and design. *Energy and Buildings,* 2003, 35, 821–841.

[230] Ouyang, J.; Gao, J.; Luo, X.; Ge, J.; Hokao, K. A study on the relationship between household lifestyles and energy consumption of residential buildings in China. *Journal of South China University of Technology,* 2007, 35, 171–174.

[231] Editorial, introduction to the special issue on energy and environment of residential buildings in China. *Energy and Buildings*, 2006, 38, 1293–1295.

[232] Ouyang, J.; Hokao, K. Energy-saving potential by improving occupants' behavior in urban residential sector in Hangzhou City, China. *Energy and Buildings,* 2009, 41, 711–720.

[233] Lopes, L.; Hokoi, S.; Miura, H.; Shuhei, K. Energy efficiency and energy savings in Japanese residential buildings—research methodology and surveyed results. *Energy and Buildings,* 2005, 37, 698–706.

[234] Ueno, T.; Inada, R.; Saeki, O.; Tsuji, K. Effectiveness of an energy-consumption information system for residential buildings. *Applied Energy,* 2006, 83, 868–883.

[235] Bohm, B.; Danig, P.O. Monitoring the energy consumption in a district heated apartment building in Copenhagen, with specific interest in the thermodynamic performance. *Energy and Buildings,* 2004, 36, 229–236.

[236] Yliniemi, K.; Delsing, J.; Deventer, J. Experimental verification of a method for estimating energy for domestic hot water production in a 2-stage district heating substation. *Energy and Buildings,* 2009, 41, 169–174.

[237] Gullev, L.; Poulsen, M. The installation of meters leads to permanent changes in consumer behaviour. *News from DBDH,* No. 3, 2006.

[238] Abrahamse, W.; Steg, L.; Vlek, C.; Rothengatter, T. A review of intervention studies aimed at household energy consumption. *Journal of Environmental Psychology,* 2005, 25, 273–291.

[239] Kempton, W.; Layne, L.L. The consumer's energy analysis environment. *Energy Policy,* 1994, 22, 857–866.

[240] Ueno, T.; Inada, R.; Saeki, O.; Tsuji, K. Effectiveness of displaying energy consumption data in residential houses: analysis of how the residents respond. In: *ECEEE 2005 Summer Study—What Works & Who Delivers,* 2005, 1289–1299.

INDEX

D

E

F

G

H

I

J

K

L

M

N

O

P

Q

R

S

T

U

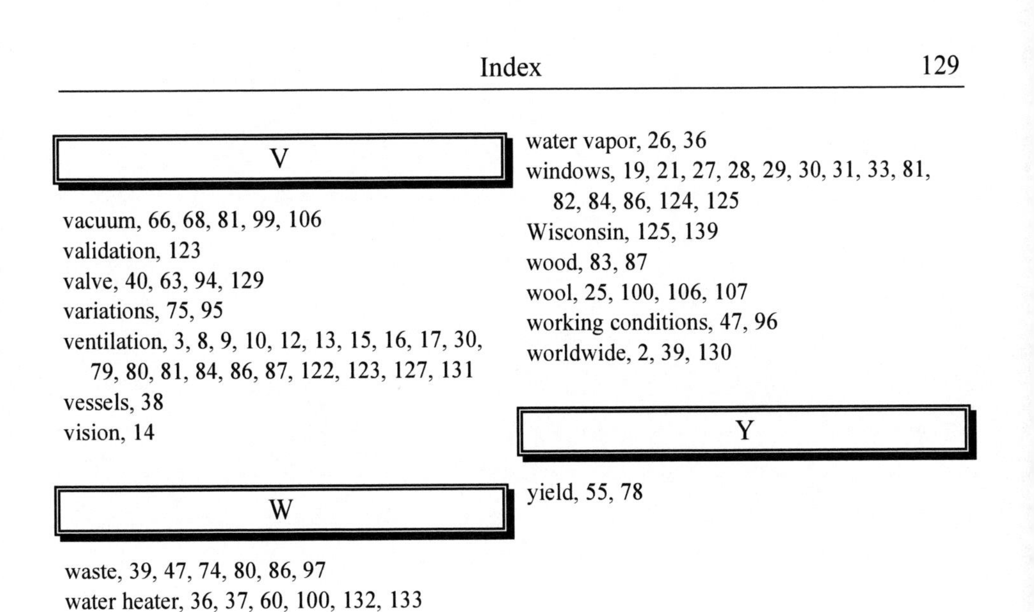

V

W

Y